AF553889

A-Z ENERGY

Prof. G.K. Bose

CENTRUM PRESS
NEW DELHI-110002 (INDIA)

CENTRUM PRESS
H.O.: 4360/4, Ansari Road, Daryaganj,
New Delhi-110 002 (India)
Ph.: 23278000, 23261597

B.O.: No. 1015, Ist Main Road, BSK IIIrd Stage
IIIrd Phase, IIIrd Block,
Bangalore - 560 085 (India)
Tel.: 080-41723429
Visit us at: www.centrumpress.com

A-Z Energy

First Edition, 2009

ISBN 978-93-80106-48-9

PRINTED IN INDIA

Printed at Salasar Imaging Systems, Delhi-110035 (India)

Contents

Preface

Energy is considered as one of the fundamental needs of human being. Its necessity arises from the role it plays in human development and economic growth. As India is generally enclosed with renewable energy sources it has to move towards greater use of renewable energy. Increasing use of these sources will be instrumental in simultaneously achieving environmental objective. Renewable energy projects in the form of solar, wind and hydropower generated electricity are the key to provide energy in rural areas where power is in short supply.

Although energy and environmental concerns were originally local in character but have now become major political issues and the subjects of international debate and regulation. Environmental degradation as a consequence of industrialization and urbanization since the last century has become a global phenomenon and it has now become a burning problem, focussing the attention of scientists, technologists, environmentalists and people in other walk of life all over the world. Several measures to preserve the environment have been suggested in international conferences held periodically and an understanding between several nations to take follow up action in adopting several measures has reached.

Author

Chapter 1

Introduction to Energy

Energy, in physical science, a term which may be defined as accumulated mechanical work, which, however, may be only partially available for use. A bent spring possesses energy, for it is capable of doing work in returning to its natural form; a charge of gunpowder possesses energy, for it is capable of doingwork in exploding; a Leyden jar charged with electricity possesses energy, for it is capable of doing work in being discharged. The motions of bodies, or of the ultimate parts of bodies, also involve energy, for stopping them would be a source of work.

All kinds of energy are ultimately measured in terms of work. If we raise of matter through a foot we do a certain amount of work against the earth's attraction; if we raise 2 lb through the same height we do twice this amount of work, and so on. Although the total energy of a system does not change with time, its value may depend on the frame of reference. For example, a seated passenger in a moving airplane has zero kinetic energy relative to the airplane, but non-zero kinetic energy relative to the earth.

ENERGY AND ITS TRANSFORMATIONS

The concept of energy and its transformations is useful in explaining and predicting most natural phenomena. The *direction* of transformations in energy is often described by entropy (equal energy spread among all available degrees of freedom) considerations, since in practice all energy transformations are permitted on a small scale, but certain larger transformations are not permitted because it is statistically unlikely that energy or matter will randomly move into more concentrated forms or smaller spaces.

The concept of energy is widespread in all sciences.

- In biology, energy is an attribute of the biological structures that is responsible for growth and development of a biological cell or an organelle of a biological organism. Energy is thus often said to be stored by cells in the structures of molecules of substances such as carbohydrates (including sugars) and lipids, which release energy when reacted with oxygen.
- In chemistry, energy is an attribute of a substance as a consequence of its atomic, molecular or aggregate structure. Since a chemical transformation is accompanied by a change in one or more of these kinds of structure, it is invariably accompanied by an increase or decrease of energy of the substances involved.
- In geology and meteorology, continental drift, mountain ranges, volcanoes, and earthquakes are phenomena that can be explained in terms of energy transformations in the Earth's interior. While meteorological phenomena like wind, rain, hail, snow, lightning, tornadoes and hurricanes, are all a result of energy transformations brought about by solar energy on the planet Earth.
- In cosmology and astronomy the phenomena of stars, nova, supernova, quasars and gamma ray bursts are the universe's highest-output energy transformations of matter. All stellar phenomena are driven by various kinds of energy transformations. Energy in such transformations is either from gravitational collapse of matter (usually molecular hydrogen) into various classes of astronomical objects (stars, black holes, etc.), or from nuclear fusion (of lighter elements, primarily hydrogen).

Energy transformations in the universe over time are characterized by various kinds of potential energy which has been available since the Big Bang, later being "released" (transformed to more active types of energy such as kinetic or radiant energy), when a triggering mechanism is available.

Familiar examples of such processes include nuclear decay, in which energy is released which was originally "stored" in heavy isotopes (such as uranium and thorium), by nucleosynthesis, a process which ultimately uses the gravitational potential energy released from the gravitational collapse of supernovae, to store energy in the creation of these heavy elements before they were incorporated into the solar system and the Earth. This energy is triggered and released in nuclear fission bombs. In a slower process, heat from nuclear decay of these atoms in the core of the Earth releases heat, which in turn may lift mountains, via orogenesis.

This slow lifting represents a kind of gravitational potential energy storage of the heat energy, which may be released to active kinetic energy in landslides, after a triggering event. Earthquakes also release stored elastic potential energy in rocks, a store which has been produced ultimately from the same radioactive heat sources. Thus, according to present understanding, familiar events such as landslides and earthquakes release energy which has been stored as potential energy in the Earth's gravitational field or elastic strain (mechanical potential energy) in rocks; but prior to this, represents energy that has been stored in heavy atoms since the collapse of long-destroyed stars created these atoms.

In another similar chain of transformations beginning at the dawn of the universe, nuclear fusion of hydrogen in the Sun releases another store of potential energy which was created at the time of the Big Bang. At that time, according to theory, space expanded and the universe cooled too rapidly for hydrogen to completely fuse into heavier elements. This meant that hydrogen represents a store of potential energy which can be released by fusion.

Such a fusion process is triggered by heat and pressure generated from gravitational collapse of hydrogen clouds when they produce stars, and some of the fusion energy is then transformed into sunlight. Such sunlight from our Sun may again be stored as gravitational potential energy after it strikes the Earth, as water evaporates from oceans and is deposited upon mountains (where, after being released at a hydroelectric dam, it can be used to drive turbine/generators to produce electricity).

Sunlight also drives many weather phenomena, save those generated by volcanic events. An example of a solar-mediated weather event is a hurricane, which occurs when large unstable areas of warm ocean, heated over months, give up some of their thermal energy suddenly to power a few days of violent air movement. Sunlight is also captured by plants as chemical potential energy, when carbon dioxide and water are converted into a combustible combination of carbohydrates, lipids, and oxygen.

Release of this energy as heat and light may be triggered suddenly by a spark, in a forest fire; or it may be available more slowly for animal or human metabolism, when these molecules are ingested, and catabolism is triggered by enzyme action. Through all of these transformation chains, potential energy stored at the time of the Big Bang is later released by intermediate events, sometimes being stored in a number of ways over time between releases, as more active energy. In all these events, one kind of energy is converted to other types of energy, including heat.

APPLICATIONS OF ENERGY

Energy is subject to a strict global conservation law; that is, whenever one measures (or calculates) the total energy of a system of particles whose interactions do not depend explicitly on time, it is found that the total energy of the system always remains constant.

- The total energy of a system can be subdivided and classified in various ways. For example, it is sometimes convenient to distinguish potential energy from kinetic energy (which is a function of coordinate time derivatives only). It may also be convenient to distinguish gravitational energy, electric energy, thermal energy, and other forms. These classifications overlap; for instance thermal energy usually consists partly of kinetic and partly of potential energy.

- The *transfer* of energy can take various forms; familiar examples include work, heat flow, and advection.
- The word "energy" is also used outside of physics in many ways, which can lead to ambiguity and inconsistency. The vernacular terminology is not consistent with technical terminology. For example, the important public-service announcement, "Please conserve energy" uses vernacular notions of "conservation" and "energy" which make sense in their own context but are utterly incompatible with the technical notions of "conservation" and "energy".

In classical physics energy is considered a scalar quantity, the canonical conjugate to time. In special relativity energy is also a scalar (although not a Lorentz scalar but a time component of the energy-momentum 4-vector). In other words, energy is invariant with respect to rotations of space, but not invariant with respect to rotations of space-time (= boosts).

Energy transfer

Because energy is strictly conserved and is also locally conserved it is important to remember that by definition of energy the transfer of energy between the "system" and adjacent regions is work. A familiar example is mechanical work. In simple cases this is written as:

$$\Delta E = W$$

if there are no other energy-transfer processes involved. Here ΔE is the amount of energy transferred, and W represents the work done on the system.

More generally, the energy transfer can be split into two categories:

$$\Delta E = W + Q$$

where Q represents the heat flow into the system.

There are other ways in which an open system can gain or lose energy. In chemical systems, energy can be added to a system by means of adding substances with different chemical potentials, which potentials are then extracted. Winding a clock would be adding energy to a mechanical system. These terms may be added to the above equation, or they can generally be subsumed into a quantity called "energy addition term E" which refers to *any* type of energy carried over the surface of a control volume or system volume.

$$\Delta E = W + Q + E$$

Where E in this general equation represents other additional advected energy terms not covered by work done on a system, or heat added to it.Energy is also transferred from potential energy (E_p) to kinetic energy (E_k) and then back to potential energy constantly. This is referred to as conservation of energy. In this closed system, energy can not be created or destroyed, so the initial energy and the final energy will be equal to each other. This can be demonstrated by the following:

$$E_{pi} + E_{ki} = E_{pF} + E_{kF}'''$$

The equation can then be simplified further since $E_p = mgh$ (mass times acceleration due to gravity times the height) and $E_k = \frac{1}{2}mv^2$ (half times mass times velocity squared).

Then the total amount of energy can be found by adding $E_p + E_k = E_{total}$.

ENERGY AND THE LAWS OF MOTION

The Hamiltonian

The total energy of a system is sometimes called the Hamiltonian, after William Rowan Hamilton. The classical equations of motion can be written in terms of the Hamiltonian, even for highly complex or abstract systems. These classical equations have remarkably direct analogs in nonrelativistic quantum mechanics.

The Lagrangian

Another energy-related concept is called the Lagrangian, after Joseph Louis Lagrange. This is even more fundamental than the Hamiltonian, and can be used to derive the equations of motion. In non-relativistic physics, the Lagrangian is the kinetic energy *minus* potential energy. Usually, the Lagrange formalism is mathematically more convenient than the Hamiltonian for non-conservative systems.

ENERGY AND THERMODYNAMICS

Internal energy

The sum of all microscopic forms of energy of a system. It is related to the molecular structure and the degree of molecular activity and may be viewed as the sum of kinetic and potential energies of the molecules; it comprises the following types of energy:

- Type Composition of Internal Energy (U)
- *Sensible energy:* the portion of the internal energy of a system associated with kinetic energies (molecular translation, rotation, and vibration; electron translation and spin; and nuclear spin) of the molecules.
- *Latent energy:* the internal energy associated with the phase of a system.
- *Chemical energy:* the internal energy associated with the different kinds of aggregation of atoms in matter.
- *Nuclear energy:* the tremendous amount of energy associated with the strong bonds within the nucleus of the atom itself.
- *Energy interactions:* Those types of energies not stored in the system (e.g. heat transfer, mass transfer, and work), but which are recognized at the system boundary as they cross it, which represent gains or losses by a system during a process.
- *Thermal energy:* The sum of sensible and latent forms of internal energy.

The Laws of Thermodynamics

According to the second law of thermodynamics, work can be totally converted into heat, but not vice versa.This is a mathematical consequence of statistical mechanics. The first law of thermodynamics simply asserts that energy is conserved, and that heat is included as a form of energy transfer. A commonly-used corollary of the first law is that for a "system" subject only to pressure forces and heat transfer (e.g. a cylinder-full of gas), the differential change in energy of the system (with a *gain* in energy signified by a positive quantity) is given by:

$$dE = TdS - PdV,$$

where the first term on the right is the heat transfer into the system, defined in terms of temperature T and entropy S (in which entropy increases and the change dS is positive when the system is heated); and the last term on the right hand side is identified as "work" done on the system, where pressure is P and volume V (the negative sign results since compression of the system requires work to be done on it and so the volume change, dV, is negative when work is done on the system).

Energy conservation in classical thermodynamics, it is highly specific, ignoring all chemical, electric, nuclear, and gravitational forces, effects such as advection of any form of energy other than heat, and because it contains a term that depends on temperature. The most general statement of the first law (i.e., conservation of energy) is valid even in situations in which temperature is undefinable.

Energy is sometimes expressed as:

$$dE = \delta Q - \delta W,$$

which is unsatisfactory because there cannot exist any thermodynamic state functions W or Q that are meaningful on the right hand side of this equation, except perhaps in trivial cases.

Equipartition of Energy

The energy of a mechanical harmonic oscillator is alternatively kinetic and potential. At two points in the oscillation cycle it is entirely kinetic, and alternatively at two other points it is entirely potential. Over the whole cycle, or over many cycles net energy is thus equally split between kinetic and potential. This is called equipartition principle - total energy of a system with many degrees of freedom is equally split among all available degrees of freedom.

This principle is vitally important to understanding the behaviour of a quantity closely related to energy, called entropy. Entropy is a measure of evenness of a distribution of energy between parts of a system. When an isolated system is given more degrees of freedom, then total energy spreads over all available degrees equally without distinction between "new" and "old" degrees. This mathematical result is called the second law of thermodynamics.

Oscillators, Phonons, and Photons

In an ensemble (connected collection) of unsynchronized oscillators, the average energy is spread equally between kinetic and potential types.In a solid, thermal energy (often referred to loosely as heat content) can be accurately described by an ensemble of thermal phonons that act as mechanical oscillators. In this model, thermal energy is equally kinetic and potential.In an ideal gas, the interaction potential between particles is essentially the delta function which stores no energy: thus, all of the thermal energy is kinetic.

Because an electric oscillator (*LC* circuit) is analogous to a mechanical oscillator, its energy must be, on average, equally kinetic and potential. It is entirely arbitrary whether the magnetic energy is considered kinetic and the electric energy considered potential, or vice versa. That is, either the inductor is analogous to the mass while the capacitor is analogous to the spring, or vice versa.

- By extension of the previous line of thought, in free space the electromagnetic field can be considered an ensemble of oscillators, meaning that radiation energy can be considered equally potential and kinetic. This model is useful, for example, when the electromagnetic Lagrangian is of primary interest and is interpreted in terms of potential and kinetic energy.
- On the other hand, in the key equation $m^2c^4 = E^2 - p^2c^2$, the contribution mc^2 is called the rest energy, and all other contributions to the energy are called kinetic energy. For a particle that has mass, this implies that the kinetic energy is $0.5p^2/ m$ at speeds much smaller

than c, as can be proved by writing $E = mc^2 \sqrt{(1 + p^2 m^{-2} c^{-2})}$ and expanding the square root to lowest order. By this line of reasoning, the energy of a photon is entirely kinetic, because the photon is massless and has no rest energy. This expression is useful, for example, when the energy-versus-momentum relationship is of primary interest.

The two analyses are entirely consistent. The electric and magnetic degrees of freedom in item 1 are transverse to the direction of motion, while the speed in item 2 is along the direction of motion. For non-relativistic particles these two notions of potential versus kinetic energy are numerically equal, so the ambiguity is harmless, but not so for relativistic particles.

Work and Virtual Work

Work is roughly force times distance. But more precisely, it is

$$W = \int F \cdot ds$$

This says that the work (W) is equal to the integral (along a certain path) of the force the mechanical work article.

Work and thus energy is frame dependent. For example, consider a ball being hit by a bat. In the centre-of-mass reference frame, the bat does no work on the ball. But, in the reference frame of the person swinging the bat, considerable work is done on the ball.

Quantum Mechanics

In quantum mechanics energy is defined in terms of the energy operator as a time derivative of the wave function. The Schrödinger equation equates the energy operator to the full energy of a particle or a system. It thus can be considered as a definition of measurement of energy in quantum mechanics. The Schrödinger equation describes the space- and time-dependence of slow changing (non-relativistic) wave function of quantum systems. The solution of this equation for bound system is discrete (a set of permitted states, each characterized by an energy level) which results in the concept of quanta. In the solution of the Schrödinger equation for any oscillator (vibrator) and for electromagnetic wave in vacuum, the resulting energy states are related to the frequency by the Planck equation $E = h\nu$(where h is the Planck's constant and í the frequency). In the case of electromagnetic wave these energy states are called quanta of light or photons.

Relativity

When calculating kinetic energy (= work to accelerate a mass from zero speed to some finite speed) relativistically - using Lorentz transformations instead of Newtonian mechanics, Einstein discovered unexpected by-product of these calculations to be an energy term which does not vanish at zero speed. He called it rest mass energy - energy which every mass must possess even when being at rest. The amount of energy is directly proportional to the mass of body:

$$E = mc^2,$$

where

m is the mass,

c is the speed of light in vacuum,

E is the rest mass energy.

For example, consider electron-positron annihilation, in which the rest mass of individual particles is destroyed, but the inertia equivalent of the system of the two particles (its invariant mass) remains (since all energy is associated with mass), and this inertia and invariant mass is carried off by photons which individually are massless, but as a system retain their mass.

This is a reversible process - the inverse process is called pair creation - in which the rest mass of particles is created from energy of two annihilating photons.

In general relativity, the *stress-energy tensor* serves as the source term for the gravitational field, in rough analogy to the way mass serves as the source term in the non-relativistic Newtonian approximation. It is not uncommon to hear that energy is "equivalent" to mass. It would be more accurate to state that every energy has inertia and gravity equivalent, and because mass is a form of energy, then mass too has inertia and gravity associated with it.

MEASUREMENT OF ENERGY

There is no absolute measure of energy, because energy is defined as the work that one system does on another. Thus, only of the transition of a system from one state into another can be defined and thus measured.The methods for the measurement of energy often deploy methods for the measurement of still more fundamental concepts of science, namely mass, distance, radiation, temperature, time, electric charge and electric current.

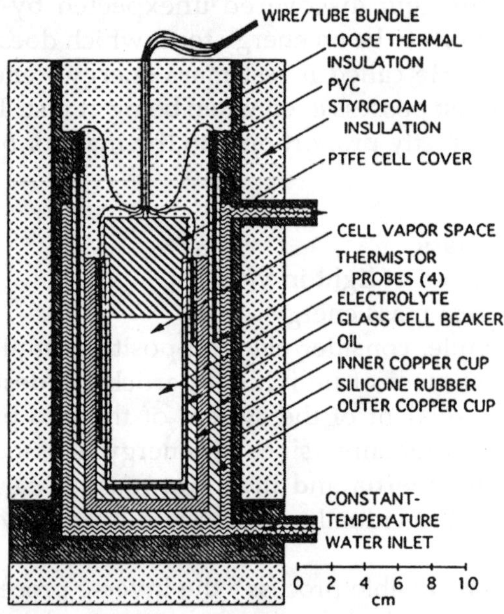

Fig. A Calorimeter

Conventionally the technique most often employed is calorimetry, a thermodynamic technique that relies on the measurement of temperature using a thermometer or of intensity of radiation using a bolometer.

Units

Throughout the history of science, energy has been

expressed in several different units such as ergs and calories. At present, the accepted unit of measurement for energy is the SI unit of energy, the joule.

FORMS OF ENERGY

Classical mechanics distinguishes between potential energy, which is a function of the position of an object, and kinetic energy, which is a function of its movement. Both position and movement are relative to a frame of reference, which must be specified: this is often (and originally) an arbitrary fixed point on the surface of the Earth, the *terrestrial* frame of reference. It has been attempted to categorize *all* forms of energy as either kinetic or potential: this is not incorrect, but neither is it clear that it is a real simplification, as Feynman points out:

These notions of potential and kinetic energy depend on a notion of length scale. For example, one can speak of *macroscopic* potential and kinetic energy, which do not include thermal potential and kinetic energy. Also what is called chemical potential energy is a macroscopic notion, and closer examination shows that it is really the sum of the potential *and kinetic* energy on the atomic and subatomic scale. Similar remarks apply to nuclear "potential" energy and most other forms of energy. This dependence on length scale is non-problematic if the various length scales are decoupled, as is often the case... but confusion can arise when different length scales are coupled, for instance when friction converts macroscopic work into microscopic thermal energy.

Potential Energy

Potential energy, symbols E_p, V or $\ddot{O}$, is defined as the work done against a given force (= work of given force with minus sign) in changing the position of an object with respect to a reference position (often taken to be infinite separation). If F is the force and s is the displacement,

$$E_p = \int F \cdot ds$$

with the dot representing the scalar product of the two vectors.

The name "potential" energy originally signified the idea that the energy could readily be transferred as work—at least in an idealized system. This is not completely true for any real system, but is often a reasonable first approximation in classical mechanics. The general equation above can be simplified in a number of common cases, notably when dealing with gravity or with elastic forces.

Gravitational Potential Energy

The gravitational force near the Earth's surface varies very little with the height, *h*, and is equal to the mass, *m*, multiplied by the gravitational acceleration, $g = 9.81\ \text{m/s}^2$. In these cases, the gravitational potential energy is given by

$$E_{p,g} = mgh$$

A more general expression for the potential energy due to Newtonian gravitation between two bodies of masses m_1 and m_2, useful in astronomy, is

$$E_{p,\,g} = -G\frac{m_1 m_2}{r},$$

where r is the separation between the two bodies and G is the gravitational constant, $6.6742(10)\times10^{-11}\ \text{m}^3\text{kg}^{-1}\text{s}^{-2}$. In this case, the reference point is the infinite separation of the two bodies.

Elastic Potential Energy

As a ball falls freely under the influence of gravity, it accelerates downward, its initial potential energy converting into kinetic energy. On impact with a hard surface the ball deforms, converting the kinetic energy into elastic potential energy. As the ball springs back, the energy converts back firstly to kinetic energy and then as the ball re-gains height into potential energy. Energy conversion to heat due to inelastic deformation and air resistance cause each successive bounce to be lower than the last.

Elastic potential energy is defined as a work needed to compress (or expand) a spring. The force, F, in a spring or any other system which obeys Hooke's law is proportional to the extension or compression, x,

$$F = -kx$$

where k is the force constant of the particular spring (or system). In this case, the calculated work becomes

$$E_{p,e} = \frac{1}{2}kx^2.$$

Hooke's law is a good approximation for behaviour of chemical bonds under normal conditions, i.e. when they are not being broken or formed.

Kinetic Energy

Kinetic energy, symbols E_k, T or K, is the work required to accelerate an object to a given speed. Indeed, calculating this work one easily obtains the following:

$$E_k = \int F \cdot dx = \int v \cdot dp = \frac{1}{2}mv^2$$

At speeds approaching the speed of light, c, this work must be calculated using Lorentz transformations, which results in the following:

$$E_k = mc^2\left(\frac{1}{\sqrt{1-(v/c)^2}} - 1\right)$$

This equation reduces to the one above it, at small (compared to c) speed. A mathematical by-product of this work (which is immediately seen in the last equation) is that even at rest a mass has the amount of energy equal to:

$$E_{rest} = mc^2$$

This energy is thus called rest mass energy.

Thermal Energy

Thermal energy (of some media - gas, plasma, solid, etc) is the energy associated with the microscopical random motion of particles constituting the media.

For example, in case of monoatomic gas it is just a kinetic energy of motion of atoms of gas as measured in the reference frame of the centre of mass of gas. In case of many-atomic gas rotational and vibrational energy is involved. In the case of liquids and solids there is also potential energy (of interaction of atoms) involved, and so on.

A heat is defined as a transfer (flow) of thermal energy across certain boundary (for example, from a hot body to cold via the area of their contact. A practical definition for small transfers of heat is

$$\Delta q = \int C_v dT$$

where C_v is the heat capacity of the system. This definition will fail if the system undergoes a phase transition—e.g. if ice is melting to water—as in these cases the system can absorb heat without increasing its temperature. In more complex systems, it is preferable to use the concept of internal energy rather than that of thermal energy.

Despite the theoretical problems, the above definition is useful in the experimental measurement of energy changes. In a wide variety of situations, it is possible to use the energy released by a system to raise the temperature of another object, e.g. a bath of water.

It is also possible to measure the amount of electric energy required to raise the temperature of the object by the same amount. The calorie was originally defined as the amount of energy required to raise the temperature of one gram of water by 1 °C (approximately 4.1855 J, although the definition later changed), and the British thermal unit was defined as the energy required to heat one pound of water by 1 °F (later fixed as 1055.06 J).

Electric Energy

The electric potential energy of given configuration of charges is defined as the work which must be done against the Coulomb force to rearrange charges from infinite separation to this configuration (or the work done by the Coulomb force separating the charges from this configuration to infinity). For two point-like charges Q_1 and Q_2 at a distance r this work, and hence electric potential energy is equal to:

$$E_{p,e} = \frac{1}{4\pi\varepsilon_0}\frac{Q_1 Q_2}{r}$$

where ε_0 is the electric constant of a vacuum, $10^7/4\pi c_0^2$ or

$8.854188...\times10^{-12}$ F/m. If the charge is accumulated in a capacitor (of capacitance *C*), the reference configuration is usually selected not to be infinite separation of charges, but vice versa - charges at an extremely close proximity to each other (so there is zero net charge on each plate of a capacitor).

The justification for this choice is purely practical - it is easier to measure both voltage difference and magnitude of charges on a capacitor plates not versus infinite separation of charges but rather versus discharged capacitor where charges return to close proximity to each other (electrons and ions recombine making the plates neutral). In this case the work and thus the electric potential energy becomes

$$E_{\text{p,e}} = \frac{Q_2}{2C}$$

If an electric current passes through a resistor, electric energy is converted to heat; if the current passes through an electric appliance, some of the electric energy will be converted into other forms of energy (although some will always be lost as heat). The amount of electric energy due to an electric current can be expressed in a number of different ways:

$$E = UQ = UIt = Pt = U^2t/\,R = I^2Rt$$

where *U* is the electric potential difference (in volts), *Q* is the charge (in coulombs), *I* is the current (in amperes), *t* is the time for which the current flows (in seconds), *P* is the power (in watts) and *R* is the electric resistance (in ohms). The last of these expressions is important in the practical measurement of energy, as potential difference, resistance and time can all be measured with considerable accuracy.

Magnetic Energy

There is no fundamental difference between magnetic energy and electric energy: the two phenomena are related by Maxwell's equations. The potential energy of a magnet of magnetic moment m in a magnetic field B is defined as the work of magnetic force (actually of magnetic torque) on re-alignment of the vector of the magnetic dipole moment, and is equal:

$$E_{\text{p,m}} = -m.\,B$$

while the energy stored in a inductor (of inductance L) when current I is passing via it is

$$E_{\mathrm{p,m}} = \frac{1}{2} LI^2.$$

This second expression forms the basis for superconducting magnetic energy storage.

Electromagnetic Fields

Calculating work needed to create an electric or magnetic field in unit volume (say, in a capacitor or an inductor) results in the electric and magnetic fields energy densities:

$$u_e = \frac{\varepsilon_0}{2} E^2$$

and

$$u_m = \frac{1}{2\mu_0} B^2$$

in SI units.

Electromagnetic radiation, such as microwaves, visible light or gamma rays, represents a flow of electromagnetic energy. Applying the above expressions to magnetic and electric components of electromagnetic field both the volumetric density and the flow of energy in e/m field can be calculated. The resulting Poynting vector, which is expressed as

$$S = \frac{1}{\mu} E \times B,$$

in SI units, gives the density of the flow of energy and its direction. The energy of electromagnetic radiation is quantized (has discrete energy levels). The spacing between these levels is equal to

$$E = h\nu$$

where h is the Planck constant, $6.6260693(11)\times10^{-34}$ Js, and *í* is the frequency of the radiation. This quantity of electromagnetic energy is usually called a photon. The photons which make up visible light have energies of 270–520 yJ, equivalent to 160–310 kJ/mol, the strength of weaker chemical bonds.

Chemical Energy

Chemical energy is the energy due to associations of atoms

in molecules and various other kinds of aggregates of matter. It may be defined as a work done by electric forces during re-arrangement of electric charges, electrons and protons, in the process of aggregation.

If the chemical energy of a system decreases during a chemical reaction, the difference is transferred to the surroundings in some form (often heat or light); on the other hand if the chemical energy of a system increases as a result of a chemical reaction - the difference then is supplied by the surroundings (usually again in form of heat or light). For example, when two hydrogen atoms react to form a dihydrogen molecule, the chemical energy *decreases* by 724 zJ (the bond energy of the H–H bond); when the electron is completely removed from a hydrogen atom, forming a hydrogen ion (in the gas phase), the chemical energy *increases* by 2.18 aJ (the ionization energy of hydrogen).

It is common to quote the changes in chemical energy for one mole of the substance in question: typical values for the change in molar chemical energy during a chemical reaction range from tens to hundreds of kJ/mol.

The chemical energy as defined above is also referred to by chemists as the internal energy, U: technically, this is measured by keeping the volume of the system constant. However, most practical chemistry is performed at constant pressure and, if the volume changes during the reaction (e.g. a gas is given off), a correction must be applied to take account of the work done by or on the atmosphere to obtain the enthalpy, H:

$$\Delta H = \Delta U + p\Delta V$$

A second correction, for the change in entropy, S, must also be performed to determine whether a chemical reaction will take place or not, giving the Gibbs free energy, G:

$$\Delta G = \Delta H - T\Delta S$$

These corrections are sometimes negligible, but often not (especially in reactions involving gases).

Since the industrial revolution, the burning of coal, oil, natural gas or products derived from them has been a socially significant transformation of chemical energy into other forms of energy. the energy "consumption" (one should really speak of "energy transformation") of a society or country is often quoted in reference to the average energy released by the combustion of these fossil fuels:

1 tonne of coal equivalent (TCE) = 29 GJ

1 tonne of oil equivalent (TOE) = 41.87 GJ

On the same basis, a tank-full of gasoline (45 litres, 12 gallons) is equivalent to about 1.6 GJ of chemical energy. Another chemically-based unit of measurement for energy is the "tonne of TNT", taken as 4.184 GJ. Hence, burning a tonne of oil releases about ten times as much energy as the explosion of one tonne of TNT: fortunately, the energy is usually released in a slower, more controlled manner.

Simple examples of chemical energy are batteries and food. When you eat the food is digested and turned into chemical energy which can be transformed to kinetic energy.

Nuclear Energy

Nuclear potential energy, along with electric potential energy, provides the energy released from nuclear fission and nuclear fusion processes. The result of both these processes are nuclei in which strong nuclear forces bind nuclear particles more strongly and closely. Weak nuclear forces (different from strong forces) provide the potential energy for certain kinds of radioactive decay, such as beta decay. The energy released in nuclear processes is so large that the relativistic change in mass (after the energy has been removed) can be as much as several parts per thousand.

Nuclear particles (nucleons) like protons and neutrons are *not* destroyed (law of conservation of baryon number) in fission and fusion processes. A few lighter particles may be created or destroyed (example: beta minus and beta plus decay, or electron capture decay), but these minor processes are not important to the immediate energy release in fission and fusion. Rather, fission and fusion release energy when collections of baryons become more tightly bound, and it is the energy associated with a fraction of the mass of the nucleons (but not the whole particles) which appears as the heat and electromagnetic radiation generated by nuclear reactions. This heat and radiation retains the "missing" mass, but the mass is missing only because it escapes in the form of heat and light, which retain the mass and conduct it out of the system where it is not measured.

The energy from the Sun, also called solar energy, is an example of this form of energy conversion. In the Sun, the process of hydrogen fusion converts about 4 million metric tons of solar matter per second into light, which is radiated into space, but during this process, the number of total protons and neutrons in the sun does not change.

In this system, the light itself retains the inertial equivalent of this mass, and indeed the mass itself (as a system), which represents 4 million tons per second of electromagnetic radiation, moving into space. Each of the helium nuclei which are formed in the process are less massive than the four protons from they were formed, but (to a good approximation), no particles or atoms are destroyed in the process of turning the sun's nuclear potential energy into light.

Surface Energy

If there is any kind of tension in a surface, such as a stretched sheet of rubber or material interfaces, it is possible to define surface energy. In particular, any meeting of dissimilar materials that don't mix will result in some kind of surface tension, if there is freedom for the surfaces to move then capillary surfaces for example, the minimum energy will as usual be sought.

A minimal surface, for example, represents the smallest possible energy that a surface can have if its energy is proportional to the area of the surface. For this reason, soap films of small size are minimal surfaces (small size reduces gravity effects, and openness prevents pressure from building up. Note that a bubble is a minimum energy surface but not a minimal surface by definition).

TRANSFORMATIONS OF ENERGY

One form of energy can often be readily transformed into

another with the help of a device- for instance, a battery, from chemical energy to electric energy; a dam: gravitational potential energy to kinetic energy of moving water (and the blades of a turbine) and ultimately to electric energy through an electric generator.

Similarly, in the case of a chemical explosion, chemical potential energy is transformed to kinetic energy and thermal energy in a very short time. Yet another example is that of a pendulum.

At its highest points the kinetic energy is zero and the gravitational potential energy is at maximum. At its lowest point the kinetic energy is at maximum and is equal to the decrease of potential energy. If one (unrealistically) assumes that there is no friction, the conversion of energy between these processes is perfect, and the pendulum will continue swinging forever.

Energy can be converted into matter and vice versa. The mass-energy equivalence formula $E = mc^2$, derived by several authors: Olinto de Pretto, Albert Einstein, Friedrich Hasenöhrl, Max Planck and Henri Poincaré, quantifies the relationship between mass and rest energy. Since c^2 is extremely large relative to ordinary human scales, the conversion of ordinary amount of mass (say, 1 kg) to other forms of energy can liberate tremendous amounts of energy ($\sim 9x10^{16}$ Joules), as can be seen in nuclear reactors and nuclear weapons.

Conversely, the mass equivalent of a unit of energy is minuscule, which is why a loss of energy from most systems is difficult to measure by weight, unless the energy loss is very large. Examples of energy transformation into matter (particles) are found in high energy nuclear physics. In nature, transformations of energy can be fundamentally classed into two kinds: those that are thermodynamically reversible, and those that are thermodynamically irreversible.

A reversible process in thermodynamics is one in which no energy is dissipated (spread) into empty energy states available in a volume, from which it cannot be recovered into more concentrated forms (fewer quantum states), without degradation of even more energy. A reversible process is one in which this sort of dissipation does not happen. For example, conversion of energy from one type of potential field to another, is reversible, as in the pendulum system described above.

In processes where heat is generated, however, quantum states of lower energy, present as possible exitations in fields between atoms, act as a reservoir for part of the energy, from which it cannot be recovered, in order to be converted with 100% efficiency into other forms of energy. In this case, the energy must partly stay as heat, and cannot be completely recovered as usable energy, except at the price of an increase in some other kind of heat-like increase in disorder in quantum states, in the universe (such as an expansion of matter, or a randomization in a crystal).

As the universe evolves in time, more and more of its energy becomes trapped in irreversible states (i.e., as heat or other kinds of increases in disorder). This has been referred to as the inevitable thermodynamic heat death of the universe.

In this heat death the energy of the universe does not change, but the fraction of energy which is available to do produce work through a heat engine, or be transformed to other usable forms of energy (through the use of generators attached to heat engines), grows less and less.

Law of Conservation of Energy

Energy is subject to the *law of conservation of energy*. According to this law, energy can neither be created (produced) nor destroyed by itself. It can only be transformed.Most kinds of energy (with gravitational energy being a notable exception) are also subject to strict local conservation laws, as well.

In this case, energy can only be exchanged between adjacent regions of space, and all observers agree as to the volumetric density of energy in any given space. There is also a global law of conservation of energy, stating that the total energy of the universe cannot change; this is a corollary of the local law, but not vice versa.

Conservation of energy is the mathematical consequence of translational symmetry of time (that is, the indistinguishability of time intervals taken at different time).

According to energy conservation law the total inflow of energy into a system must equal the total outflow of energy from the system, plus the change in the energy contained within the system.

This law is a fundamental principle of physics. It follows from the translational symmetry of time, a property of most phenomena below the cosmic scale that makes them independent of their locations on the time coordinate. Put differently, yesterday, today, and tomorrow are physically indistinguishable. Thus is because energy is the quantity which is canonical conjugate to time.

This mathematical entanglement of energy and time also results in the uncertainty principle - it is impossible to define the exact amount of energy during any definite time interval. The uncertainty principle should not bc confused with energy conservation - rather it provides mathematical limits to which energy can in principle be defined and measured.

In quantum mechanics energy is expressed using the Hamiltonian operator. On any time scales, the uncertainty in the energy is by

$$\Delta E \Delta t \geq \frac{\hbar}{2}$$

which is similar in form to the Heisenberg uncertainty principle (but not really mathematically equivalent thereto, since H and t are not dynamically conjugate variables, neither in classical nor in quantum mechanics).

In particle physics, this inequality permits a qualitative understanding of virtual particles which carry momentum, exchange by which and with real particles, is responsible for the creation of all known fundamental forces (more accurately known as fundamental interactions).

Virtual photons (which are simply lowest quantum mechanical energy state of photons) are also responsible for electrostatic interaction between electric charges (which results in Coulomb law), for spontaneous radiative decay of exited atomic and nuclear states, for the Casimir force, for van der Waals bond forces and some other observable phenomena.

Energy and Life

Any living organism relies on an external source of energy—radiation from the Sun in the case of green plants; chemical energy in some form in the case of animals—to be able to grow and reproduce.

The daily 1500–2000 Calories (6^{-8} MJ) recommended for a human adult are taken as a combination of oxygen and food molecules, the latter mostly carbohydrates and fats, of which glucose ($C_6H_{12}O_6$) and stearin ($C_{57}H_{110}O_6$) are convenient examples. The food molecules are oxidised to carbon dioxide and water in the mitochondria

$$C_6H_{12}O_6 + 6O_2 \rightarrow 6CO_2 + 6H_2O$$

$$C_{57}H_{110}O_6 + 81.5O_2 \rightarrow 57CO_2 + 55H_2O$$

and some of the energy is used to convert ADP into ATP

$$ADP + HPO_4^{2''} \rightarrow ATP + H_2O$$

The rest of the chemical energy in the carbohydrate or fat is converted into heat: the ATP is used as a sort of "energy currency", and some of the chemical energy it contains when split and reacted with water, is used for other metabolism (at each stage of a metabolic pathway, some chemical energy is converted into heat). Only a tiny fraction of the original chemical energy is used for work:

- Gain in kinetic energy of a sprinter during a 100 m race: 4 kJ
- Gain in gravitational potential energy of a 150 kg weight lifted through 2 metres: 3kJ
- Daily food intake of a normal adult: 6^{-8} MJ

It would appear that living organisms are remarkably inefficient (in the physical sense) in their use of the energy they receive (chemical energy or radiation), and it is true that most real machines manage higher efficiencies.

However, in growing organisms the energy that is converted to heat serves a vital purpose, as it allows the organism tissue to be highly ordered with regard to the molecules it is built from.

The second law of thermodynamics states that energy (and matter) tends to become more evenly spread out across the universe: to concentrate energy (or matter) in one specific place, it is necessary to spread out a greater amount of energy (as heat) across the remainder of the universe ("the surroundings"). Simpler organisms can achieve higher energy efficiencies than more complex ones, but the complex organisms can occupy ecological niches that are not available to their simpler brethren.

The conversion of a portion of the chemical energy to heat at each step in a metabolic pathway is the physical reason behind the pyramid of biomass observed in ecology: to take just the first step in the food chain, of the estimated 124.7 Pg/a of carbon that is fixed by photosynthesis, 64.3 Pg/a (52%) are used for the metabolism of green plants, i.e. reconverted into carbon dioxide and heat.

Chapter 2

Electric Current

Electricity, one of the basic forms of energy. Electricity is associated with electric charge, a property of certain elementary particles such as electrons and protons, two of the basic particles that make up the atoms of all ordinary matter. Electric charges can be stationary, as in static electricity, or moving, as in an electric current.

Electrical activity takes place constantly everywhere in the universe. Electrical forces hold molecules together. The nervous systems of animals work by means of weak electric signals transmitted between neurons. Electricity is generated, transmitted, and converted into heat, light, motion, and other forms of energy through natural processes, as well as by devices built by people.

Electricity is an extremely versatile form of energy. It can be generated in many ways and from many different sources. It can be sent almost instantaneously over long distances. Electricity can also be converted efficiently into other forms of energy, and it can be stored. Because of this versatility, electricity plays a part in nearly every aspect of modern technology. Electricity provides light, heat, and mechanical power. It makes telephones, computers, televisions, and countless other necessities and luxuries possible.

ELECTRIC CHARGE

Electricity consists of charges carried by electrons, protons, and other particles. Electric charge comes in two forms: positive and negative. Electrons and protons both carry exactly the same amount of electric charge, but the positive charge of the proton is exactly opposite the negative charge of the electron. If an object has more protons than electrons, it is said to be positively charged; if it has more electrons than protons, it is said to be negatively charged. If an object contains as many protons as electrons, the charges will cancel each other and the object is said to be uncharged, or electrically neutral. Electricity occurs in two forms: static electricity and electric current. Static electricity consists of electric charges that stay in one place. An electric current is a flow of electric charges between objects or locations.

STATIC ELECTRICITY

Static electricity can be produced by rubbing together two objects made of different materials. Electrons move from the surface of one object to the surface of the other if the second material holds onto its electrons more strongly than the first does. The object that gains electrons becomes negatively charged, since it now has more electrons than protons. The object that gives up electrons becomes positively charged. For example, if a nylon comb is run through clean, dry hair, some of the electrons on the hair are transferred to the comb. The comb becomes negatively charged and the hair becomes positively charged. The following materials are named in decreasing order of their ability to hold electrons: rubber, silk, glass, flannel, and fur. If any two of these materials are rubbed together, the material earlier in the list becomes negative, and the material later in the list becomes positive. The materials should be clean and dry.

Charging by Contact

Objects become electrically charged in either of two ways: by contact or by induction.

A charged object transfers electric charge to an object with lesser charge if the two touch. When this happens, a charge flows from the first to the second object for a brief time. Charges in motion form an electric current. When charge flows between objects in contact, the amount of charge that an object receives depends on its ability to store charge. The ability to store charge is called capacitance and is measured in units called farads.

Charging by contact can be demonstrated by touching an uncharged electroscope with a charged comb. An electroscope is a device that contains two strips of metal foil, called leaves, that hang from one end of a metal rod. A metal ball is at the other end of the rod. When the charged comb touches the ball, some of the charges on the comb flow to the leaves, which separate because they now hold like charges and repel each other. If the comb is removed, the leaves remain apart because they retain their charges. The electroscope has thus been charged by contact with the comb.

This flow of charge between objects with different amounts of charge will occur whenever possible. However, it requires a pathway for the electric charge to move along. Some materials, called conductors, allow an electric current to flow through them easily. Other materials, called insulators, strongly resist the passage of an electric current.

Under normal conditions, air is an insulator. However, if an object gains a large enough charge of static electricity, part of the charge may jump, or discharge, through the air to another object without touching it directly. When the charge is large enough, the air becomes a conductor. Lightning is an example of a discharge.

Coulomb's Law

Objects with opposite charges attract each other, and objects with similar charges repel each other. Coulomb's law, formulated by French physicist Charles Augustin de Coulomb during the late 18th century, quantifies the strength of the attraction or repulsion. This law states that the force between two charged objects is directly proportional to the product of their charges and inversely proportional to the square of the distance between them. The greater the charges on the objects, the larger the force between them; the greater the distance between the objects, the lesser the force between them. The unit of electric charge, also named after Coulomb, is equal to the combined charges of 6.24×10^{18} protons (or electrons).

If two charged objects in contact have the same capacitance, they divide the charge evenly. Suppose, for example, that one object has a charge of +4 coulombs and the other a charge of +8 coulombs. When they touch, charge will flow from the 8-coulomb object to the 4-coulomb object until each has a charge of +6 coulombs. If each object originally had a charge of +6 coulombs, no charge would flow between them.

If two objects have different capacitances, they divide the charge in proportion to their capacitances. If an object with a capacitance of 10 farads touches an object with a capacitance of 5 farads, the 10-farad object will end up with twice the amount of charge of the 5-farad object. Suppose that the objects are oppositely charged and that one has a charge of +20 coulombs and the other a charge of -8 coulombs. Their total charge is therefore +12 coulombs. After they touch, the 10-farad object will have a charge of +8 coulombs and the 5-farad object will have +4 coulombs.

Charging by Induction

A charged object may induce a charge in a nearby neutral object without touching it. For example, if a positively charged object is brought near a neutral object, the electrons in the neutral object are attracted to the positive object. Some of these electrons flow to the side of the neutral object that is nearest to the positive object.

This side of the neutral object accumulates electrons and becomes negatively charged. Because electrons leave the far side of the neutral object while its protons remain stationary, that side becomes positively charged. Since the negatively charged side of the neutral object is closest to the positive object, the attraction between this side and the positive object is greater than the repulsion between the positively charged side and the positive object. The net effect is an attraction between the objects. Similarly, when a negatively charged object is brought near a neutral object, the negative object induces a positive charge on the near side of the neutral object and a negative charge on the far side. As before, the net effect is an attraction between the objects.

The induced charges described above are not permanent. As soon as the charged object is taken away, the electrons on the other object redistribute themselves evenly over it, so that it again becomes neutral.

An object can also be charged permanently by induction. If a negatively charged object, A, is brought near a neutral object, B, the electrons on B are repelled as far as possible from A and flow to the other side of B. If that side of B is then connected to the ground by a good conductor, such as a metal wire, the electrons flow out through the wire into the ground. The ground can receive almost any amount of charge because Earth, being neutral, has an enormous capacitance. Object B is said to be grounded by the wire connecting it to Earth.

If this wire is then removed, B has a positive charge, since it has lost electrons to Earth. Thus B has been permanently charged by induction. Even if A is subsequently removed, B still remains positive because the wire has been disconnected and B cannot regain electrons from Earth to neutralize its positive charge.

NELECTRIC CURRENT

An electric current is a movement of charge. When two objects with different charges touch and redistribute their charges, an electric current flows from one object to the other until the charge is distributed according to the capacitances of the objects. If two objects are connected by a material that lets charge flow easily, such as a copper wire, then an electric current flows from one object to the other through the wire. Electric current can be demonstrated by connecting a small light bulb to an electric battery by two copper wires. When the connections are properly made, current flows through the wires and the bulb, causing the bulb to glow.

Current that flows in one direction only, such as the current in a battery-powered flashlight, is called direct current. Current that flows back and forth, reversing direction again and again, is called alternating current. Direct current, which is used in most battery-powered devices, is easier to understand than alternating current. Alternating current, which is used in most devices that are "plugged in" to electrical outlets in buildings.

Other properties that are used to quantify and compare electric currents are the voltage (also called electromotive force) driving the current and the resistance of the conductor to the passage of the current. The amount of current, voltage, and resistance in any circuit are all related through an equation called Ohm's law.

Conductors and Insulators

Conductors are materials that allow an electric current to flow through them easily. Most metals are good conductors.

Substances that do not allow electric current to flow through them are called insulators, nonconductors, or dielectrics. Rubber, glass, and air are common insulators. Electricians wear rubber gloves so that electric current will not pass from electrical equipment to their bodies. However, if an object contains a sufficient amount of charge, the charge can arc, or jump, through an insulator to another object. For example, if you shuffle across a wool rug and then hold your finger very close to, but not in contact with, a metal doorknob or radiator, current will arc through the air from your finger to the doorknob or radiator, even though air is an insulator. In the dark, the passage of the current through the air is visible as a tiny spark.

Measuring Electric Current

Electric current is measured in units called amperes (amp). If 1 coulomb of charge flows past each point of a wire every second, the wire is carrying a current of 1 amp. If 2 coulombs flow past each point in a second, the current is 2 amp.

Voltage

When the two terminals of a battery are connected by a conductor, an electric current flows through the conductor. One terminal continuously sends electrons into the conductor, while the other continuously receives electrons from it. The current flow is caused by the voltage, or potential difference, between the terminals. The more willing the terminals are to give up and receive electrons, the higher the voltage. Voltage is measured in units called volts. Another name for a voltage produced by a source of electric current is electromotive force.

Resistance

A conductor allows an electric current to flow through it, but it does not permit the current to flow with perfect freedom. Collisions between the electrons and the atoms of the conductor interfere with the flow of electrons. This phenomenon is known as resistance. Resistance is measured in units called ohms. The symbol for ohms is the Greek letter omega, Ω.

A good conductor is one that has low resistance. A good insulator has a very high resistance. At commonly encountered temperatures, silver is the best conductor and copper is the second best. Electric wires are usually made of copper, which is less expensive than silver.

The resistance of a piece of wire depends on its length, and its cross-sectional area, or thickness. The longer the wire is, the greater its resistance. If one wire is twice as long as a wire of identical diameter and material, the longer wire offers twice as much resistance as the shorter one. A thicker wire, however, has less resistance, because a thick wire offers more room for an electric current to pass through than a thin wire does. A wire whose cross-sectional area is twice that of another wire of equal length and similar material has only half the resistance of the thinner wire. Scientists describe this relationship between resistance, length, and area by saying that resistance is proportional to length and inversely proportional to cross-sectional area. Usually, the higher the temperature of a wire, the greater its resistance. The resistance of some materials drops to zero at very low temperatures. This phenomenon is known as superconductivity.

Ohm's Law

The relationship between current, voltage, and resistance is given by Ohm's law. This law states that the amount of current passing through a conductor is directly proportional to the voltage across the conductor and inversely proportional to the resistance of the conductor. Ohm's law can be expressed as an equation, $V = IR$, where V is the difference in volts between two locations (called the potential difference), I is the amount of current in amperes that is flowing between these two points, and R is the resistance in ohms of the conductor between the two locations of interest. $V = IR$ can also be written $R = V/I$ and $I = V/R$. If any two of the quantities are known, the third can be calculated. For example, if a potential difference of 110 volts sends a 10-amp current through a conductor, then the resistance of the conductor is $R = V/I = 110/10 = 11$ ohms. If $V = 110$ and $R = 11$, then $I = V/R = 110/11 = 10$ amp.

Under normal conditions, resistance is constant in conductors made of metal. If the voltage is raised to 220 in the example above, then R is still 11. The current I will be doubled, however, since $I = V/R = 220/11 = 20$ amp.

Heat and Power

A conductor's resistance to electric current produces heat. The greater the current passing through the conductor, the greater the heat. Also, the greater the resistance, the greater the heat. A current of I amp passing through a resistance of R ohms for t seconds generates an amount of heat equal to I^2Rt joules (a joule is a unit of energy equal to 0.239 calorie).

Energy is required to drive an electric current through a resistance. This energy is supplied by the source of the current, such as a battery or an electric generator. The rate at which energy is supplied to a device is called power, and it is often measured in units called watts. The power P supplied by a current of I amp passing through a resistance of R ohms is given by $P = I^2R$.

HOW ELECTRIC CURRENT IS CONDUCTED

All electric currents consist of charges in motion. However, electric current is conducted differently in solids, gases, and liquids. When an electric current flows in a solid conductor, the flow is in one direction only, because the current is carried entirely by electrons. In liquids and gases, however, a two-directional flow is made possible by the process of ionizationConduction in Solids

The conduction of electric currents in solid substances is made possible by the presence of *free electrons* (electrons that are free to move about). Most of the electrons in a bar of copper, for example, are tightly bound to individual copper atoms. However, some are free to move from atom to atom, enabling current to flow.

Ordinarily the motion of the free electrons is random; that is, as many of them are moving in one direction as in another. However, if a voltage is applied to the two ends of a copper bar by means of a battery, the free electrons tend to drift toward one end. This end is said to be at a higher potential and is called the positive end.

The other end is said to be at a lower potential and is called the negative end. The function of a battery or other source of electric current is to maintain potential difference. A battery does this by supplying electrons to the negative end of the bar to replace those that drift to the positive end and also by absorbing electrons at the positive end.

Insulators cannot conduct electric currents because all their electrons are tightly bound to their atoms. A perfect insulator would allow no charge to be forced through it, but no such substance is known at room temperature. The best insulators offer high but not infinite resistance at room temperature.

Some substances that ordinarily have no free electrons, such as silicon and germanium, can conduct electric currents when small amounts of certain impurities are added to them. Such substances are called semiconductors. Semiconductors generally have a higher resistance to the flow of current than does a conductor, such as copper, but a lower resistance than an insulator, such as glass.

Conduction in Gases

Gases normally contain few free electrons and are generally insulators. When a strong potential difference is applied between two points inside a container filled with a gas, the few free electrons are accelerated by the potential difference and collide with the atoms of the gas, knocking free more electrons. The gas atoms become positively charged ions and the gas is said to be ionized. The electrons move toward the high-potential (more positive) point, while the ions move toward the low-potential (more negative) point. An electric current in a gas is composed of these opposite flows of charges.

Conduction in Liquid Solutions

Many substances become ionized when they dissolve in water or in some other liquid. An example is ordinary table salt, sodium chloride (NaCl). When sodium chloride dissolves in water, it separates into positive sodium ions, Na^+, and negative chlorine ions, Cl^-. If two points in the solution are at different potentials, the negative ions drift toward the positive point, while the positive ions drift toward the negative point. As in gases, the electric current is composed of these flows of opposite charges. Thus, while water that is absolutely pure is an insulator, water that contains even a slight impurity of an ionized substance is a conductor.

Since the positive and negative ions of a dissolved substance migrate to different points when an electric current flows, the substance is gradually separated into two parts. This separation is called electrolysis.

SOURCES OF ELECTRIC CURRENT

There are several different devices that can supply the voltage necessary to generate an electric current. The two most common sources are generators and electrolytic cells.

Generators

Generators use mechanical energy, such as water pouring through a dam or the motion of a turbine driven by steam, to produce electricity. The electric outlets on the walls of homes and other buildings, from which electricity to operate lights and appliances is drawn, are connected to giant generators located in electric power stations. Each outlet contains two terminals. The voltage between the terminals drives an electric current through the appliance that is plugged into the outlet.

Electrolytic Cells

Electrolytic cells use chemical energy to produce electricity. Chemical reactions within an electrolytic cell produce a potential difference between the cell's terminals. An electric battery consists of a cell or group of cells connected together.

Other Sources

There are many sources of electric current other than generators and electrolytic cells. Fuel cells, for example, produce electricity through chemical reactions. Unlike electrolytic cells, however, fuel cells do not store chemicals and therefore must be constantly refilled.

Certain sources of electric current operate on the principle that some metals hold onto their electrons more strongly than other metals do. Platinum, for example, holds its electrons less strongly than aluminum does. If a strip of platinum and a strip of aluminum are pressed together under the proper conditions, some electrons will flow from the platinum to the aluminum. As the aluminum gains electrons and becomes negative, the platinum loses electrons and becomes positive.

The strength with which a metal holds its electrons varies with temperature. If two strips of different metals are joined and the joint heated, electrons will pass from one strip to the other. Electricity produced directly by heating is called thermoelectricity. Some substances emit electrons when they are struck by light. Electricity produced in this way is called photoelectricity. When pressure is applied to certain crystals, a potential difference develops across them. Electricity thus produced is called piezoelectricity. Some microphones work on this principle.

ELECTRIC CIRCUITS

An electric circuit is an arrangement of electric current sources and conducting paths through which a current can continuously flow. In a simple circuit consisting of a small light bulb, a battery, and two pieces of wire, the electric current flows from the negative terminal of the battery, through one piece of connecting wire, through the bulb filament (also a type of wire), through the other piece of connecting wire, and back to the positive terminal of the battery. When the electric current flows through the filament, the filament heats up and the bulb lights.

A switch can be placed in one of the connecting wires. A flashlight is an example of such a circuit. When the switch is open, the connection is broken, electric current cannot flow through the circuit, and the bulb does not light. When the switch is closed, current flows and the bulb lights.

The bulb filament may burn out if too much electric current flows through it. To prevent this from happening, a fuse (circuit breaker) may be placed in the circuit. When too much current flows through the fuse, a wire in the fuse heats up and melts, thereby breaking the circuit and stopping the flow of current. The wire in the fuse is designed to melt before the filament would melt.

The part of an electric circuit other than the source of electric current is called the load. The load includes all appliances placed in the circuit, such as lights, radios, fans, buzzers, and toasters. It also includes the connecting wires, as well as switches, fuses, and other devices. The load forms a continuous conducting path between the terminals of the current source.

There are two basic ways in which the parts of a circuit are arranged. One arrangement is called a series circuit, and the other is called a parallel circuit.

Series Circuits

If various objects are arranged to form a single conducting path between the terminals of a source of electric current, the objects are said to be connected in series. The electron current first passes from the negative terminal of the source into the first object, then flows through the other objects one after another, and finally returns to the positive terminal of the source. The current is the same throughout the circuit. In the example of the light bulb, the wires, bulb, switch, and fuse are connected in series.

When objects are connected in series, the electric current flows through them against the resistance of the first object, then against the resistance of the next object, and so on. Therefore the total resistance to the current is equal to the sum of the individual resistances.

If three objects with resistances R_1, R_2, and R_3 are connected in series, their total resistance is $R_1 + R_2 + R_3$. For example, if a motor with a resistance of 48 ohms is connected to the terminals of a current source by two wires, each with a resistance of 1 ohm, the total resistance of the motor and wires is $48 + 1 + 1 = 50$ ohms. If the voltage is 100 volts, a current of $100/50 = 2$ amp will flow through the circuit.

Voltage can be thought of as being used up by the objects in a circuit. The voltage that each object uses up is called the voltage drop across that object. Voltage drop can be calculated from the equation $V = IR$, where V is the voltage drop across the object, I is the amount of current, and R is the resistance of the object.

In the example of the motor, the voltage drop in each wire is $V = IR = 2 \times 1 = 2$ volts, and the voltage drop in the motor is $2 \times 48 = 96$ volts. Adding up the voltage drops $(2 + 2 + 96)$ gives a total drop of 100 volts. In a series circuit the sum of the voltage drops across the objects always equals the total voltage supplied by the source.

Parallel Circuits

If various objects are connected to form separate paths between the terminals of a source of electric current, they are said to be connected in parallel. Each separate path is called a branch of the circuit. Current from the source splits up and enters the various branches. After flowing through the separate branches, the current merges again before reentering the current source.

The total resistance of objects connected in parallel is less than that of any of the individual resistances. This is because a parallel circuit offers more than one branch (path) for the electric current, whereas a series circuit has only one path for all the current.

The electric current through a parallel circuit is distributed among the branches according to the resistances of the branches. If each branch has the same resistance, then the current in each will be equal. If the branches have different resistances, the current in each branch can be determined from the equation $I = V/R$, where I is the amount of current in the branch, V is the voltage, and R is the resistance of the branch.

The total resistance of a parallel circuit can be calculated from the equation

$$\frac{1}{R} = \frac{1}{R_1} + \frac{1}{R_2} + \frac{1}{R_3} + \cdots$$

where R is the total resistance and R_1, R_2,... are the resistances of the branches. For example, if a parallel circuit consists of three branches with resistances of 10, 15, and 30 ohms, then

$$\frac{1}{R} = \frac{1}{10} + \frac{1}{15} + \frac{1}{30} = \frac{3}{30} + \frac{2}{30} + \frac{1}{30} = \frac{6}{30} = \frac{1}{5}$$

Therefore, $R = 5$ ohms. In this circuit, a voltage of 150 volts would produce an electric current of $I = V/R = 150/5 = 30$ amp.

The greater the resistance of a given branch, the smaller the portion of the electric current flowing through that branch. If a parallel circuit of three branches, with resistances of 10, 15, and 30 ohms, is connected to a 150-volt source, the branch with a resistance of 10 ohms would receive a current of $V/R = 150/10 = 15$ amp. Similarly, the 15-ohm branch receives 10 amp, and the 30-ohm branch receives 5 amp. These branch currents add up to a total current of 30 amp, which is the value obtained by dividing the voltage by the total resistance.

Series-Parallel Circuits

Many circuits combine series and parallel arrangements. One branch of a parallel circuit, for example, may have within it several objects in a series. The resistances of these objects must be combined according to the rules for a series circuit.

On the other hand, a series circuit may at one point divide into two or more branches and then rejoin.

The branches are parallel and must be treated by the rules for parallel circuits. Complicated series-parallel circuits may be analyzed by means of two rules called Kirchhoff's laws. These rules make it possible to find the amount of electric current flowing through each part of any circuit, as well as the voltage across it.

The first of Kirchhoff's laws states that at any junction in a circuit through which a steady current is flowing, the sum of the currents flowing to the junction is equal to the sum of the currents flowing away from that point. The second law states that, starting at any point in a circuit and following any closed path back to the starting point, the net sum of the voltage encountered will be equal to the net sum of the products of the resistances encountered and the currents flowing through them. In other words, Ohm's law applies not only to a circuit as a whole.

Series and Parallel Sources

Sources of electric current can also be connected in various ways. Sources can be arranged in series by connecting a terminal of one source to the opposite terminal of the next source. For example, if the positive terminal of battery A is connected to the negative terminal of battery B, and the positive terminal of battery B to the negative terminal of battery C, then batteries A, B, and C are in series. The load is then placed between the positive terminal of battery C and the negative terminal of battery A.

When sources of electric current are connected in series, their total voltage is equal to the sum of their individual voltages. For example, three 1.5-volt batteries connected in series furnish a total of 4.5 volts. If the load is 9 ohms, the batteries send a current of 4.5/9 = 0.5 amp through the load. Current sources may be arranged in parallel by connecting all the positive terminals together and all the negative terminals together. The load is then placed between the group of positive terminals and the group of negative terminals.

Arranging sources in parallel does not increase the voltage. If three 1.5-volt batteries are connected in parallel, the total voltage is still 1.5 volts. Batteries should not be connected in parallel unless they have approximately the same voltage. If a high voltage battery is connected in parallel with a low voltage battery, the high voltage battery will force an electric current through the low voltage battery and damage it.

ELECTRIC FIELDS

A single electric charge can attract or repel, and it will demonstrate this ability as soon as another charge is brought near it. The ability to attract or repel can be thought of as being stored in the region around the charge. This region is called the electric field of force of the charge. All charged objects have electric fields around them.

Lines of Force

An electric field can be visualized as consisting of imaginary lines called lines of force. Each line corresponds to the path that a positive charge would take if placed in the field on that line. The lines in the field around a positively charged object radiate in all directions away from the object, since the object repels positive charges. Conversely, the lines in the field around a negatively charged object are directed toward the object. If a positive and a negative object are placed near each other, their lines of force connect. If two objects with similar charges are placed near each other, the lines do not connect. Lines of force never cross each other. Lines of force are only imaginary. Nevertheless, the idea of lines of force helps in visualizing an electric field.

Field Direction

When a charge is placed at any given point in an electric field, it is acted on by a force that tends to push it in a certain direction. This direction is called the direction of the field at that point. The field direction can be represented graphically by the lines of force near an electric charge.

Field Strength

The strength, or intensity, of a field at any point is defined as the force exerted on a charge of 1 coulomb placed at that point. For example, if a point charge of 1 coulomb is subjected to a force of 10 newtons, the electric field is 10 newtons per coulomb at that point. An object with a charge of 5 coulombs would be subjected to a force of 50 newtons at the same point.

Field strength is represented graphically by the closeness (density) of the lines of force. Where the lines are close together, the field is strong. Where they are far apart, the field is weak. Near a charge, the field is strong and the lines are close together. At greater distances from the charge, the field weakens and the lines are not as close together. The field strength values that the lines represent are relative, since a field can be drawn with as many lines as desired.

ELECTRICITY AND MAGNETISM

Many similarities exist between electric and magnetic phenomena. A magnet has two opposite poles, referred to as north and south. Opposite magnetic poles attract each other, and similar magnetic poles repel each other, exactly as happens with electric charges.

The force with which magnetic poles attract or repel each other depends on the strength of the poles and the distance between them. This relationship is similar to the Coulomb's inverse square law for electric charges. The similarities between electric and magnetic phenomena indicate that electricity and magnetism are related. Electricity produces magnetic effects and magnetism produces electric effects. The relationship between electricity and magnetism is called electromagnetism.

Magnetic Effects of Electricity

It has been noted that an electric field exists around any electric charge. If electric charges are moving, they constitute an electric current. The magnetic effect of electricity is demonstrated by the fact that a magnetic field exists around any electric current. The field can be detected when a magnet is brought close to the current-carrying conductor.

The magnetic field around an electric current can be

thought of as lines of magnetic force that form closed circular loops around the wire that carries the current. The direction of the magnetic field can be determined by a convenient rule called the right-hand rule. To apply this rule, the thumb of the right hand is pointed in the direction in which the current is flowing and the fingers are curled around the wire. The direction of the fingers then indicates the direction of the lines of magnetic force. (The right-hand rule assumes that current flows from positive to negative.)

Motor Effect

As already stated, a magnetic field exists around a wire carrying an electric current, and a magnetic field exists between the two poles of a magnet. If the wire is placed between the poles, the magnetic fields interact to produce a force that tends to push the wire out of the field. This phenomenon, known as the motor effect, is used in electric motors.

Solenoids

If a wire is bent into many continuous loops to form a long spiral coil, then the magnetic lines of force tend to go through the centre of the coil from one end to the other rather than around the individual loops of wire. Such a coil, called a solenoid, behaves in the same way as a magnet and is the basis for all electromagnets. The end from which the lines exit is the north pole and the end into which the lines reenter is the south pole. The polarity of the coil can be determined by applying the left-hand coil rule. If the left hand grasps the coil in such a way that the fingers curl around in the direction of the electron current, then the thumb points in the direction of the north pole.

Electric Effects of Magnetism

If a wire is moved through a magnetic field in such a way that it cuts the magnetic lines of force, a voltage is created across the wire. An electric current will flow through the wire if the two ends of the wire are connected by a conductor to form a circuit. This current is called an induced current, and the induction of a current in this manner is called electromagnetic induction.

It does not matter whether the wire moves or the magnetic field moves, provided that the wire cuts through lines of force. If a magnet is moved near a stationary wire, the lines of magnetic force are cut by the wire and an electric current is induced in the wire.

Like any electric current, an induced current generates a magnetic field around it. Lenz's law expresses an important fact concerning this magnetic field: The motion of an induced current is always in such a direction that its magnetic field opposes the magnetic field that is causing the current.

Alternating Current

An alternating current is an electric current that changes direction at regular intervals. When a conductor is moved back and forth in a magnetic field, the flow of current in the conductor will reverse direction as often as the physical motion of the conductor reverses direction. Most electric power stations supply electricity in the form of alternating currents. The current flows first in one direction, builds up to a maximum in that direction, and dies down to zero.

It then immediately starts flowing in the opposite direction, builds up to a maximum in that direction, and again dies down to zero. Then it immediately starts in the first direction again. This surging back and forth can occur at a very rapid rate. Two consecutive surges, one in each direction, are called a cycle. The number of cycles completed by an electric current in one second is called the frequency of the current. In the United States and Canada, most currents have a frequency of 60 cycles per second.

Although direct and alternating currents share some characteristics, some properties of alternating current are somewhat different from those of direct current. Alternating currents also produce phenomena that direct currents do not. Some of the unique traits of alternating current make it ideal for power generation, transmission, and use.

Amperage and Voltage

The strength, or amperage, of an alternating current varies continuously between zero and a maximum. Since it is inconvenient to take into account a whole range of amperage values, scientists simply deal with the effective amperage. Like a direct current, an alternating current produces heat as it passes through a conductor. The effective amperage of an alternating current is equal to the amperage of a direct current that produces heat at the same rate. In other words, 1 effective amp of alternating current through a conductor produces heat at the same rate as 1 amp of direct current flowing through the same conductor. Similarly, the voltage of an alternating current is considered in terms of the effective voltage.

Impedance

Like direct current, alternating current is hindered by the resistance of the conductor through which it passes. In addition, however, various effects produced by the alternating current itself hinder the alternating current. These effects depend on the frequency of the current and on the design of the circuit, and together they are called reactance. The total hindering effect on an alternating current is called impedance. It is equal to the resistance plus the reactance.

The relationship of effective current, effective voltage, and impedance is expressed by $V = IZ$, where V is the effective voltage in volts, I is the effective current in amperes (amp), and Z is the impedance in ohms.

Advantages of Alternating Current

Alternating current has several characteristics that make it more attractive than direct current as a source of electric power, both for industrial installations and in the home. The most important of these characteristics is that the voltage or the current may be changed to almost any value desired by means of a simple electromagnetic device called a transformer. When an alternating current surges back and forth through a coil of wire, the magnetic field about the coil expands and collapses and then expands in a field of opposite polarity and again collapses. In a transformer, a coil of wire is placed in the magnetic field of the first coil, but not in direct electric connection with it. The movement of the magnetic field induces an alternating current in the second coil.

If the second coil has more turns than the first, the voltage induced in the second coil will be larger than the voltage in the first, because the field is acting on a greater number of individual conductors. Conversely, if there are fewer turns in the second coil, the secondary, or induced, voltage will be smaller than the primary voltage.

The action of a transformer makes possible the economical transmission of electric power over long distances. If 200,000 watts of power is supplied to a power line, it may be equally well supplied by a potential of 200,000 volts and a current of 1 amp or by a potential of 2,000 volts and a current of 100 amp, because power is equal to the product of voltage and current. The power lost in the line through heating, however, is equal to the square of the current times the resistance. Thus, if the resistance of the line is 10 ohms, the loss on the 200,000-volt line will be 10 watts, whereas the loss on the 2,000-volt line will be 100,000 watts, or half the available power. Accordingly, power companies tend to favour high voltage lines for long distance transmission.

HISTORY

Humans have known about the existence of static electricity for thousands of years, but scientists did not make great progress in understanding electricity until the 1700s.

Early Theories

The ancient Greeks observed that amber, when rubbed, attracted small, light objects. About 600 BC Greek philosopher Thales of Miletus held that amber had a soul, since it could make other objects move. In a treatise written about three centuries later, another Greek philosopher, Theophrastus, stated that other substances also have this power. For almost 2,000 years after Theophrastus, little progress was made in the study of electricity. He gave these substances the Latin name *electrica,* which is derived from the Greek word *elektron* (which means "amber"). The word *electricity* was first used by English writer and physician Sir Thomas Browne in 1646.

The fact that electricity can flow through a substance was discovered by 17th-century German physicist Otto von Guericke, who observed conduction in a linen thread. Von Guericke also described the first machine for producing an electric charge in 1672. The machine consisted of a sulfur sphere turned by a crank.

When a hand was held against the sphere, a charge was induced on the sphere. Conduction was rediscovered independently by Englishman Stephen Gray during the early 1700s. Gray also noted that some substances are good conductors while others are insulators. Also during the early 1700s, Frenchman Charles Dufay observed that electric charges are of two kinds. He found that opposite kinds attract each other while similar kinds repel. Dufay called one kind vitreous and the other kind resinous.

American scientist Benjamin Franklin theorized that electricity is a kind of fluid. According to Franklin's theory, when two objects are rubbed together, electric fluid flows from one object to the other. The object that gains electric fluid acquires a vitreous charge, which Franklin called positive charge. The object that loses electric fluid acquires a resinous charge, which Franklin called negative charge.

Franklin demonstrated that lightning is a form of electricity. In 1752 he constructed a kite and flew it during a storm. When the string became wet enough to conduct, Franklin, who stood under a shed and held the string by a dry silk cord, put his hand near a metal key attached to the string. A spark jumped. Electric charge gathered by the kite had flowed down the wet string to the key and then jumped across an air gap to flow to the ground through Franklin's body. Franklin also showed that a Leyden jar, a device able to store electric charge, could be charged by touching it to the key when electric current was flowing down the string.

Around 1766 British chemist Joseph Priestley proved experimentally that the force between electric charges varies inversely with the square of the distance between the charges. Priestley also demonstrated that an electric charge distributes itself uniformly over the surface of a hollow metal sphere and that no charge and no electric field of force exists within such a sphere.

French physicist Charles Augustin de Coulomb reinvented a torsion balance to measure accurately the force exerted by electric charges. With this apparatus he confirmed Priestley's observations and also showed that the force between two charges is proportional to the product of the individual charges. In 1791 Italian biologist Luigi Galvani published the results of experiments that he had performed on the muscles of dead frogs. Galvani had found earlier that the muscles in a frog's leg would contract if he applied an electric current to them.

19th and 20th Centuries

In 1800 another Italian scientist, Alessandro Volta, announced that he had created the voltaic pile, a form of electric battery. The voltaic pile made the study of electric current much easier by providing a reliable, steady source of current. Danish physicist Hans Christian Oersted demonstrated that electric currents are surrounded by magnetic fields in 1819.

Shortly afterward, André Marie Ampère discovered the relationship known as Ampere's law, which gives the direction of the magnetic field. Ampère also demonstrated the magnetic properties of solenoids. Georg Simon Ohm, a German high school teacher, investigated the conducting abilities of various metals. In 1827 Ohm published his results, including the relationship now known as Ohm's law.

In 1830 American physicist Joseph Henry discovered that a moving magnetic field induces an electric current. The same effect was discovered a year later by English scientist Michael Faraday. Faraday introduced the concept of lines of force, a concept that proved extremely useful in the study of electricity.

About 1840 British physicist James Prescott Joule and German scientist Hermann Ludwig Ferdinand von Helmholtz demonstrated that electricity is a form of energy and that electric circuits obey the law of the conservation of energy.

Also during the 19th century, British physicist James Clerk Maxwell investigated the properties of electromagnetic waves and light and developed the theory that the two are identical. Maxwell summed up almost all the laws of electricity and magnetism in four mathematical equations. His work paved the way for German physicist

Heinrich Rudolf Hertz, who produced and detected electric waves in the atmosphere in 1886, and for Italian engineer Guglielmo Marconi, who harnessed these waves in 1895 to produce the first practical radio signaling system.

The electron theory, which is the basis of modern electrical theory, was first advanced by Dutch physicist Hendrik Antoon Lorentz in 1892.

American physicist Robert Andrews Millikan accurately measured the charge on the electron in 1909. The widespread use of electricity as a source of power is largely due to the work of pioneering American engineers and inventors such as Thomas Alva Edison, Nikola Tesla, and Charles Proteus Steinmetz during the late 19th and early 20th centuries.

Chapter 3

Electric Power Transmission

The transport of generator-produced electric energy to loads. An electric power transmission system interconnects generators and loads and generally provides multiple paths among them. Multiple paths increase system reliability because the failure of one line does not cause a system failure. Most transmission lines operate with three-phase alternating current (ac). The standard frequency in North America is 60 Hz; in Europe, 50 Hz. The three-phase system has three sets of phase conductors. Long-distance energy transmission occasionally uses high-voltage direct-current (dc) lines.

The electric power system can be divided into the distribution, subtransmission, and transmission systems. With operating voltages less than 34.5 kV, the distribution system carries energy from the local substation to individual households, using both overhead and underground lines. With operating voltages of 69-138 kV, the subtransmission system distributes energy within an entire district and regularly uses overhead lines.

With operating voltage exceeding 230 kV, the transmission system interconnects generating stations and large substations located close to load centers by using overhead lines. In the early days of commercial use of electric power, transmission of electric power at the same voltage as used by lighting and mechanical loads restricted the distance between generating plant and consumers.

In 1882 generation was with direct current, which could not easily be increased in voltage for long-distance transmission. Different classes of loads, for example, lighting, fixed motors, and traction (railway) systems, required different voltages and so used different generators and circuits.

The so-called "universal system" used transformers both to couple generators to high-voltage transmission lines, and to connect transmission to local distribution circuits. By a suitable choice of utility frequency, both lighting and motor loads could be served. Rotary converters and later mercury-arc valves and other rectifier equipment allowed DC load to be served by local conversion where needed. Even generating stations and loads using different frequencies could also be interconnected using rotary converters.

By using common generating plants for every type of load, important economies of scale were achieved, lower overall capital investment was required, load factor on each plant was increased allowing for higher efficiency, allowing for a lower cost of energy to the consumer and increased overall use of electric power.

By allowing multiple generating plants to be interconnected over a wide area, electricity production cost was reduced. The most efficient available plants could be used to supply the varying loads during the day. Reliability was improved and capital investment cost was reduced, since stand-by generating capacity could be shared over many more customers and a wider geographic area. Remote and low-cost sources of energy, such as hydroelectric power or mine-mouth coal, could be exploited to lower energy production cost.

The first transmission of three-phase alternating current using high voltage took place in 1891 during the international electricity exhibition in Frankfurt. A 25 kV transmission line, approximately 175 kilometers long, connected Lauffen on the Neckar and Frankfurt.

Voltages used for electric power transmission increased throughout the 20th century. By 1914 fifty-five transmission systems operating at more than 70,000 V were in service, the highest voltage then used was 150,000 volts.

The rapid industrialization in the 20th century made electrical transmission lines and grids a critical part of the economic infrastructure in most industrialized nations.

Interconnection of local generation plants and small distribution networks was greatly spurred by the requirements of World War I, where large electrical generating plants were built by governments to provide power to munitions factories; later these plants were connected to supply civil load through long-distance transmission.

ALTERNATING-CURRENT TRANSMISSION

Overhead transmission lines distribute the majority of the electric energy in the system. A typical high-voltage line has three phase conductors to carry the current and transport the energy, and two grounded shield conductors to protect the line from direct lightning strikes.

The usually bare conductors are insulated from the supporting towers by insulators attached to grounded towers or poles. Lower-voltage lines use post insulators, while the high-voltage lines are built with insulator chains or long-rod composite insulators. The normal distance between the supporting towers is a few hundred feet.

Transmission lines use ACSR (aluminum cable, steel reinforced) and ACAR (aluminum cable, alloy reinforced) conductors. In an ACSR conductor, a stranded steel core carries the mechanical load, and layers of stranded aluminum surrounding the core carry the current. An ACAR conductor is a stranded cable made of an aluminum alloy with low resistance and high mechanical strength.

ACSR conductors are usually used for high-voltage lines, and ACAR conductors for subtransmission and distribution lines. Ultrahigh-voltage (UHV) and extrahigh-voltage (EHV) lines use bundle conductors. Each phase of the line is built with two, three, or four conductors connected in parallel and separated by about 1.5 ft (0.5 m). Bundle conductors reduce corona discharge.

Transmission lines are subject to environmental adversities, including wide variations of temperature, high winds, and ice and snow deposits. Typically designed to withstand environmental stresses occurring once every 50–100 years, lines are intended to operate safely in adverse conditions.Variable weather affects line operation. Extreme weather reduces corona inception voltage, leading to an increase in audible noise, radio noise, and telephone interference. Load variation requires regulation of line voltage. A short circuit generates large currents, overheating conductors and producing permanent damage.

The power that a line can transport is limited by the line's electrical parameters. Voltage drop is the most important factor for distribution lines; where the line is supplied from only one end, the permitted voltage drop is about 5%.Conductor temperature must be lower than the temperature which causes permanent elongation.

A typical maximum steady-state value for ACSR is 212°F (100°C), but in an emergency temperatures 10–20% higher are allowed for a short period of time (10 min to 1 h).

Corona discharge is generated when the electric field at the surface of the conductor becomes larger than the breakdown strength of the air. The oscillatory nature of the discharge generates high-frequency, short-duration current pulses, the source of corona-generated radio and television interference. Surface irregularities such as water droplets cause local field concentration, enhancing corona generation. Thus, during bad weather, corona discharge is more intense and losses are much greater. Corona discharge also generates audible noise with two components: a broad-band, high-frequency component, which produces crackling and hissing, and a 120-Hz pure tone.

Transmission-line conductors are surrounded by an electric field which decreases as distance from the line increases, and depends on line voltage and geometry. At ground level, this field induces current and voltage in grounded bodies, causes corona in grounded objects, and can induce fuel ignition. Utilities limit the electric field at the perimeter of right-of-ways to about 1000 V/m. An ac magnetic field around the transmission line also decreases with distance from the line.

Lightning strikes produce high voltages and traveling waves on transmission lines, causing insulator flashovers and interruption of operation. Steel grounded shield conductors at the tops of the towers significantly reduce, but do not eliminate, the probability of direct lightning strikes to phase conductors.The operation of circuit breakers causes switching surges that can result in interruption of inductive current, energization of lines with trapped charges, and single-phase ground fault. Modern circuit breakers, operating in two steps, reduce switching surges to 1.5–2 times the 60-Hz voltage.

Line current induces a disturbing voltage in telephone lines running parallel to transmission lines. Because the induced voltage depends on the mutual inductance between the two lines, disturbance can be reduced by increasing the distance between the lines and shielding the telephone lines.

Underground Power Transmission

Most cities use underground cables to distribute electrical energy. These cables virtually eliminate negative environmental effects and reduce electrocution hazards. However, they entail significantly higher construction costs.Underground cables are divided into two categories: distribution cables (less than 69 kV) and high-voltage power-transmission cables (69–500 kV).Extruded solid dielectric cables dominate in the 15–33-kV urban distribution system. In a typical arrangement, the stranded copper or aluminum conductor is shielded by a semiconductor layer, which reduces the electric stress on the conductor's surface.

Oil-impregnated paper-insulated distribution cables are used for higher voltages and in older installations.Cable temperatures vary with load changes, and cyclic thermal expansion and contraction may produce voids in the cable. High voltage initiates corona in the voids, gradually destroying cable insulation. Low-pressure oil-filled cable construction reduces void formation.

A single-phase concentric cable has a hollow conductor with a central oil channel. Three-phase cables have three oil channels located in the filler. Electric power can also be transmitted by underground power cables instead of overhead power lines. This is a more expensive option, as the life-cycle cost of an underground power cable is two to four times the cost of an overhead power line. However, they can assist the transmission of power across:

- Densely populated urban areas
- Areas where land is unavailable or planning consent is difficult
- Rivers and other natural obstacles
- Land with outstanding natural or environmental heritage
- Areas of significant or prestigious infrastructural development
- Land whose value must be maintained for future urban expansion and rural development

Compared to overhead lines, underground cables emit much less powerful magnetic fields. (All conductors carrying current which varies with respect to time generate magnetic fields.) Underground cables need a narrower strip of about 1-10 metres to install, whereas the lack of cable insulation requires an overhead line to be installed on a strip of about 20- 200 metres wide to be kept permanently clear for safety, maintenance and repair. Those advantages can in some cases justify the higher investment cost.

Most high-voltage underground cables for power transmission that are currently sold on the market are insulated by a sheath of cross linked polyethylene (XLPE). Some cable may have a lead jacket in conjunction with XLPE insulation to allow for fibre optics to be seamlessly integrated within the cable. Before 1960, underground power cables used to be insulated with oil and paper and ran in a rigid steel pipe, or a semi-rigid aluminium or lead jacket or sheath.

The oil was kept under pressure to prevent formation of voids that would allow partial discharges within the cable insulation. There are still many of those oil-and-paper insulated cables in use worldwide· Between 1960 and 1990, polymers became more widely used at distribution voltages, mostly EPDM; however, their relative unreliability - particularly early XLPE - resulted in a slow uptake at transmission voltages.

- *Submarine cables:* High-voltage cables are frequently used for crossing large bodies of water. Water provides natural cooling, and pressure reduces the possibility of void formation. A typical submarine cable has cross-linked polyethylene insulation, and corrosion-resistant aluminum alloy wire armoring that provides tensile strength and permits installation in deep water.
- *Transmission lines:* A system of conductors suitable for conducting electric power or signals between two or more termini. Transmission lines take many forms in practice and have application in many disciplines. For example, they traverse the countryside, carrying telephone signals and electric power. The same transmission lines, with similar functions, may be hidden above false ceilings in urban buildings. With the need to reliably and securely transmit ever larger amounts of data, the required frequency of operation has increased from the high-frequency microwave range to the still higher frequency of light. Optical fibers are installed in data-intensive buildings and form a nationwide network. Increasing demand also requires that transmission lines handle greater values of electric power.

  Transmission lines can, in some cases, be analyzed by using a fairly simple model that consists of distributed linear electrical components. Models of this type, with some permutations, can also be used to describe wave propagation in integrated circuits and along nerve fibers in animals. The study of hollow metal waveguides or optical fibers is usually based upon an analysis starting from Maxwell's equations rather than employing transmission-line models. Fundamentals and definitions that can initially be obtained from a circuit model of a transmission line carry over to waveguides, where the analysis is more complicated.
- *Coaxial cables and strip lines:* Two particular types of transmission lines for communication that have

received considerable attention are the coaxial cable and the strip line. The coaxial cable is a flexible transmission line and typically is used to connect two electronic instruments together. In the coaxial cable, a dielectric separates a centre conducting wire from a concentric conducting sleeve. The strip line is used in integrated circuits to connect, say, two transistor circuits together.

A strip line also has a dielectric that separates the top conducting element from the base, which may be an electrical ground plane in the circuit.

- *Circuit model:* These transmission lines can be most easily analyzed in terms of electrical circuit elements consisting of distributed linear inductors and capacitors. The values of these elements are in terms of the physical dimensions of the coaxial cable and the strip line, and the permittivity of the dielectric. Each of the elements is interpreted to be measured in terms of a unit length of the element. An equivalent circuit represents either the coaxial cable or the strip line as well as other transmission lines such as two parallel wires. Losses are incorporated into the transmission-line model with the addition of a distributed resistance in series with the inductor and a distributed conductance in parallel with the capacitor. Additional distributed circuit elements can be incorporated into the model in order to describe additional effects. For example, the linear capacitors could be replaced with reverse-biased varactor diodes and the propagation of nonlinear solitons could be studied.
- *Power transmission lines:* Electric power generating stations and load consumption centers are connected by a network of power transmission lines, mostly overhead lines. Power transmitted is generally in the form of three-phase alternating current (ac) at 60 or 50 Hz. In a few instances, where a clear technical or economic advantage exists, direct-current (dc) systems may be used. As the distances over which the power

must be transmitted become great and as the amount of power transmitted increases, the power lost in the transmission lines becomes an important component of the production cost of electricity, and it becomes advantageous to increase the transmission voltage. This basic consideration has led to electric power networks which use higher voltages for long-distance bulk power transfers, with several layers of underlying regional networks at progressively lower voltages which extend over shorter distances. The most common transmission voltages in use are 765, 500, 400, 220 kV, and so forth. Voltages below 69 kV are termed subtransmission or distribution voltages, and at these and lower voltages the networks may have fewer alternative supply paths (loops) or may be entirely radial in structure.

NATURE OF ELECTRICITY

Electricity is generated as it is used. Unlike other commodities, there is very little ability to store electricity. Because of the instantaneous nature of the electric system, constant modifications must be made to assure that the generation of power matches the consumption of power. The electric system we've grown to depend on is very complex and dynamic, ever adjusting to meet changing needs.

The amount of power on a transmission line at any given moment depends on generation production and dispatch, customer use, the status of other transmission lines and their associated equipment, and even the weather. The transmission system must accommodate changing electricity supply and demand conditions, unexpected outages, planned shutdowns of generators or transmission equipment for maintenance, weather extremes, fuel shortages, and other challenges.

The Transmission Grid

The electrical transmission system is more complex and dynamic than other utility systems, such as water or natural gas. Electricity flows from power plants, through transformers and transmission lines, to substations, distribution lines, and then finally to the electricity consumer. The electric system is highly interconnected. The interconnectedness of the system means that the transmission grid functions as one entity. Power entering the system flows along all available paths, not just from Point A to Point B.

The system does not recognize divisions between service areas, counties, states, or even countries. The current transmission grid includes not only transmission lines that run from power plants to load centers, but also from transmission line to transmission line, providing a redundant system that helps assure the smooth flow of power.

If a transmission line is taken out of service in one part of the power grid, the power can usually be rerouted through other power lines to continue delivering electricity to the customer. In essence, the electricity from many power plants is "pooled" in the transmission system, and each distribution system draws from this pool.

This networked system helps to achieve a high reliability for power delivery, since any one power plant that shuts down should only constitute a fraction of the power being delivered by the power grid. This pooling of power means that power is provided from a diversity of sources, including coal, nuclear, natural gas, oil, or other renewable energy sources such as hydropower, biomass, wind, or solar power.

Transmission Outages

A transmission line outage acts like a dam, forcing the electricity around the blockage onto other lines. If adjacent transmission lines cannot handle the power that is rerouted, safety devices may switch them off, further impeding power flows and potentially leading to cascading outages and system failure, i.e., a blackout.

This is one of the disadvantages of the interconnectedness of the transmission grid. Failures in one location can quickly affect the entire system, producing a large scale blackout. For reliable power transmission, a region requires backup transmission lines with adequate capacity.

Components of the Transmission System

Power plants generate three-phase alternating current (AC). This means that there are three wires coming out of every plant.

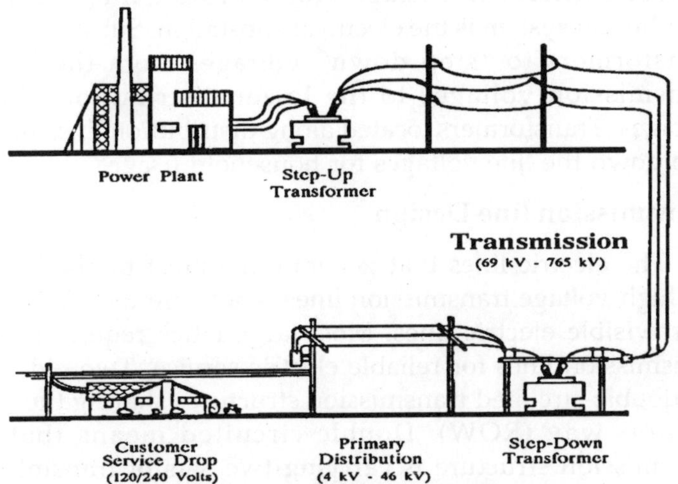

Fig. Simplified Electric System

On a transmission structure, the three large wires are called conductors and carry the electric power. They are usually about an inch in diameter. There is also a smaller wire at the top of the structure, called a shield wire. The shield wire is designed to protect the power line from lightning. Poles with two sets of three wires (conductors) are called "double-circuit" poles. Sometimes a distribution line is strung under the transmission lines reducing the need for additional power poles.

Electricity is transferred from the power plant to the users, through the electric grid. The grid consists of two separate infrastructures: the high-voltage transmission system and the lower-voltage distribution system. High-voltage transmission lines minimize electrical losses and are therefore used to carry electricity hundreds of miles. Transmission lines in Wisconsin range from 69 to 345 kilovolts (kV). Higher voltage lines, such as 500 and 765 kV, are not used in Wisconsin but are in other states. The lower-voltage lines (distribution system) draw electricity from the transmission lines and distribute it to individual customers.

Lower voltage lines range from 12 to 24 kV. The voltage that connects to your house is 120 to 240 volts. The interface between different voltage transmission lines and the distribution system is the electrical substation. Substations use transformers to "step down" voltages from the higher transmission voltages to the lower distribution system voltages. Transformers located along distribution lines further step down the line voltages for household usage.

Transmission line Design

The electric lines that generate the most public interest are high-voltage transmission lines. These are the largest and most visible electric lines. Most large cities require several transmission lines for reliable electric service. Two older 345 kV double-circuited transmission structures sharing the same rightof- way (ROW). Double-circuited means that the transmission structure is carrying two sets of transmission lines, each with three conductors.

Fig. Two Older Double-Circuit Transmission Structures

Transmission lines are larger than the more common distribution lines that exist along rural roads and city streets. Transmission line poles or structures are between 60 and 140 feet tall. Distribution line structures are approximately 40 feet tall. There are several different kinds of transmission structures. Transmission structures can be constructed of metal or wood. They can be single-poled or double poled. They can be single-circuited carrying one set of transmission lines or double-circuited with two sets of lines. A close up of a commonly built double-circuited, single-pole transmission structure. The different types of transmission structures.

Fig. Typical Double-Circuit Transmission Structures

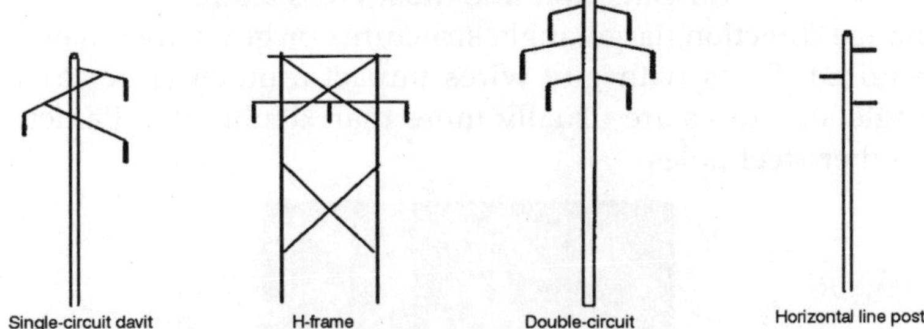

Fig. Different Transmission Structures

Different transmission structures have different material and construction costs, and require different rightof- way widths, distances between structures (span length), and pole height. These issues also vary with different voltages. In the past, many transmission lines were constructed on H-frame wood structures and metal lattice structures.

New lines are most often constructed with single pole structures because of rightof- way width limitations and environmental considerations. Current right-of-way widths vary between 80 to 140 feet. Pole height and load capacity limitations control allowable span length either on the basis of ground clearance or ability to support heavy wind and ice loads. In areas where single-pole structures are preferred, weak or wet soils may require concrete foundations for support.

Fig. Two Angle Structures and a Transmission Line Crossing a Road

Where a transmission line must cross a street or slightly change direction, large angle structures or guy wires may be required. Poles with guy wires impact a much larger area. Angle structures are usually more than double the diameter of other steel poles.

Fig. Close-up of the Base of an Angle Structure

They are made of steel, usually five to six feet in diameter, and have a large concrete base. The base may be buried ten or more feet below the ground surface. The diameter of the pole and the depth the base is buried depends on the condition of the soils and the voltage of the line.

Wisconsin's Transmission System

There are currently nearly 11,500 miles of transmission line in Wisconsin. The transmission lines in Wisconsin which are over 100 kV. High-voltage transmission lines deliver large amounts of power on a regional basis. The higher the voltage, the more power the line can carry. The Wisconsin transmission system has a general electric flow from northwest to southeast through the state. The western part of Wisconsin is connected by high-voltage lines (161 and 345 kV) primarily from Minnesota. The southeastern part of Wisconsin is connected to northern Illinois by 345 kV high voltage lines.

However, there are few connections between these two parts of the state and a lack of redundant connections. Redundant connects are needed to ensure reliable service. The Wisconsin transmission system is currently congested under normal power flow conditions. In addition there are many transmission lines in Wisconsin that are more than 60 years old, requiring upgrades or replacement. Some areas, due to the age of the lines and the lack of redundant lines, often experience disruptions in electrical power.

Restructuring the electric industry and Wisconsin transmission: In the recent past, individual utilities owned power plants, substations, transmission lines, and distribution lines that generated and provided electricity to their customers. This system was known as verticallyintegrated.

Restructuring the electric industry in Wisconsin resulted in the transfer of much of the state's transmission infrastructure to a separate company. Transmission lines in the western portion of the state are owned by Xcel Energy Services Inc (Xcel) and Dairyland Power Cooperative (DPC).Both Xcel and Dairyland have facilities in other states and still own transmission and generation facilities.

Wisconsin Transmission Needs

There are several drivers for new transmission construction:

- Growth in an area's electricity use, which often requires new distribution substations and new lines to connect them to the existing transmission system, or increased capacity on existing transmission lines. Wisconsin has a growth in electrical demand of about 2 percent per year.
- Older transmission lines that are in poor condition and no longer reliable might need to be replaced by new lines. Often new lines will have a greater electricity carrying capacity than those they replace.
- The existing transmission system must be reinforced with new lines to prevent equipment overloads and low voltages.
- New power plants need new transmission lines to connect them to the existing transmission system. Transmission line projects approved by the Public Service Commission of Wisconsin (PSC) are required by state statute. They must satisfy the reasonable needs of the public for adequate electric energy supply and must not be overbuilt or be designed in excess of probable future electric needs.

Regulation of the Electric Industry

The PSC regulates the siting and construction of new transmission lines. The Electric Reliability Organization (ERO) sets and enforces reliability standards. ERO replaced the organization formerly known as NERC, North American Electric Reliability Council, a voluntary utility-managed reliability organization. The FERC oversees the ERO. Wisconsin utilities belong to two different organizations under the ERO umbrella: Reliability First and the Midwest Reliability Organization.

Regional Transmission Organizations

In 1992, the Energy Policy Act (EPAct) opened access to

the transmission grid for non-utility power generators. As ordered by the EPAct, FERC passed orders that required owners of transmission lines to provide open, non-discriminatory access to power generators engaged in interstate commerce.

These policies allowed for wholesale competition—that is, competition among power generators for contracts with utilities, as well as contracts with large industrial firms. During the years 1996 to 2000, FERC ordered utilities to offer other energy providers fair and open access to their transmission lines. FERC created RTOs to provide independent oversight over the nation's power grid. Wisconsin is part of the Midwest Independent System Operator (MISO).

Community Planning

In prior decades, electric transmission lines were constructed from Point A to Point B in the most direct manner possible with limited concern for communities, crops, natural resources, or private property issues. As these older lines require improvements, they may be rerouted to share corridors with roads, and to avoid, where practicable, community and natural resource impacts. At the same time, continued growth in energy usage will require new electric substations and transmission lines to be sited and constructed.

New and upgraded electric facilities may impact many communities and many property owners. To meet future growth, communities often draft plans for sewers, roads, and development districts, but few cities, towns, or counties include transmission lines in their plans. Transmission lines are costly to build and difficult to site. Cities, towns, and counties can help reduce land use conflicts by:

- Dedicating a strip of land along existing transmission corridors for potential future right-of-way expansions,
- Identifying potential transmission corridors and substation sites in new developments, and
- Defining set-backs or lot sizes for properties adjacent to transmission lines so that buildings don't constrain future use of the right-of-ways.

Being an active participant in the decision-making process will improve the ability of communities to manage future growth and protect their resources.

Advanced Transmission Technologies

Not all new power transmission technologies are currently ready for commercial use. Many are still in the experimental and prototype stage. The new technologies mostly fall into two categories – new materials that may increase the amount of power that can be safely transferred through right-of-ways, and devices that more finely control the flow of power.

New power control devises improve the capacity of existing lines. New material advances include high-temperature super-conducting technology and advanced composite conductors. New technologies that better manage the flow of power include high-voltage direct current systems, superconducting magnetic energy storage, flexible alternating current transmission system devices, and real-time ratings on transmission lines. The disadvantage of many of these new technologies is that they are still being researched and their cost is extremely high.

High-Temperature Superconducting (HTSC) technology

The conductors in HTSC devices operate at extremely low resistances. However, they require refrigeration (generally liquid nitrogen) to super-cool the conductors, increasing maintenance costs and the complexity of the system. The benefits are cables that can carry five times as much power as traditional copper wires with the same dimensions. This greatly reduces the number of new transmission lines and the amount of new right-of-way required.

Composite Conductors

Usually transmission lines contain steel-core cables that support strands of aluminum wires which are the primary conductors of electricity. New cores developed from composite materials reduce sagging with high temperatures associated with more power going through transmission lines. Life-cycle costs of the experimental conductors are high. The installation and maintenance procedures continue to be developed.

Superconducting Magnetic Energy Storage

SMES devices would be strategically located in a transmission grid to damp out disturbances. SMES systems use a cryogenic technology to store energy by circulating current in a super-conducting coil, advanced line-monitoring equipment to detect voltage deviations, and inverters that can rapidly inject the appropriate combination of real and reactive power to counteract voltage problems.

By correcting for potential stability problems, these systems permit the operation of transmission lines at capacities much closer to their thermal limits than currently possible. However the expense of these cooling systems is a disadvantage. American Transmission Company currently has six SMES devices in use on its system to address low-voltage and grid instability issues.

Flexible AC Transmission System Devices

Currently, the transmission grid relies mostly on slow, electromechanical switches and human operators. Mechanical switching creates split-second delays in responding to problems. A variety of power-electronic power switching and other devices are being developed that will operate much faster, along with sensors capable of spotting disturbances instantly. These systems will improve control and stability of the transmission grid and permit transmission lines to operate closer to their thermal limits, making better use of existing wires. However these FACTS devices are very expensive, costing several hundred million dollars for the conversion of a single line.

Real-time Ratings of Transmission Lines

This is another use of advanced information technologies to expand the capacity of existing transmission systems. Special devices can measure the real time tension in transmission lines, ambient temperature and wind speed, or cable sag. The results of the measurements are telemetered to the control centre which then adjusts the line rating accordingly. Once again the drawback with this technology is the high cost relative to the incremental potential increase in capacity.

Bulk Power Transmission

Engineers design transmission networks to transport the energy as efficiently as feasible, while at the same time taking into account economic factors, network safety and redundancy. These networks use components such as power lines, cables, circuit breakers, switches and transformers.

A transmission substation decreases the voltage of electricity coming in allowing it to connect from long distance, high voltage transmission, to local, lower voltage, distribution. It also reroutes power to other transmission lines that serve local markets. A transmission substation may include phase-shifting or voltage regulating transformers.

Transmission efficiency is improved by increasing the voltage using a step-up transformer, which reduces the current in the conductors, while keeping the power transmitted nearly equal to the power input. The reduced current flowing through the conductor reduces the losses in the conductor and since, according to Joule's Law, the losses are proportional to the square of the current, halving the current makes the transmission loss one quarter the original value.

A transmission grid is a network of power stations, transmission circuits, and substations. Energy is usually transmitted within the grid with three-phase AC. DC systems require relatively costly conversion equipment which may be economically justified for particular projects. Single phase AC is used only for distribution to end users since it is not usable for large polyphase induction motors.

In the 19th century two-phase transmission was used, but required either three wires with unequal currents or four wires. Higher order phase systems require more than three wires, but deliver marginal benefits. The capital cost of electric power stations is so high, and electric demand is so variable, that it is often cheaper to import some portion of the variable load than to generate it locally. Because nearby loads are often correlated, imported electricity must often come from far away. Because of the economics of load balancing, transmission grids now span across countries and even large portions of continents. The web of interconnections between power producers and consumers ensures that power can flow even if a few links are inoperative.

The unvarying (or slowly varying over many hours) portion of the electric demand is known as the "base load", and is generally served best by large facilities (and therefore efficient due to economies of scale) with low variable costs for fuel and operations, i.e. nuclear, coal, hydro. Renewables such as solar, wind, ocean/tidal, etc. are not considered "base load" but can still add power to the grid. Smaller- and higher-cost sources such as combined cycle or combustion turbine plants that run on natural gas are then added as needed.

Long-distance transmission of electricity (thousands of miles) is cheap and efficient. Thus distant suppliers can be cheaper than local sources. Multiple local sources (even if more expensive and infrequently used) can make the transmission grid more fault tolerant to weather and other disasters that can disconnect distant suppliers.

Long distance transmission allows remote renewable energy resources to be used to displace fossil fuel consumption. Hydro and wind sources can't be moved closer to high population cities, and solar costs are lowest in remote areas where local power needs are the least. Connection costs alone can determine whether any particular renewable alternative is economically sensible. Costs can be prohibitive for transmission lines.

- *Grid input:* At the generating plants the energy is produced at a relatively low voltage between about 2300 volts and 30,000 volts, depending on the size of the unit. The generator terminal voltage is then stepped up by the power station transformer to a higher voltage for transmission over long distances.
- *Losses:* Transmitting electricity at high voltage reduces

the fraction of energy lost to Joule heating. For a given amount of power, a higher voltage reduces the current and thus the resistive losses in the conductor. For example, raising the voltage by a factor of 10 reduces the current by a corresponding factor of 10 and therefore the I^2R losses by a factor of 100, provided the same sized conductors are used in both cases. Even if the conductor size is reduced x10 to match the lower current I^2R the losses are still reduced x10. Long distance transmission is typically done with overhead lines at voltages of 115 to 1,200 kV. At extremely high voltages, more than 2,000 kV between conductor and ground, corona discharge losses are so large that they can offset the lower resistance loss in the line conductors.

In an alternating current circuit, the inductance and capacitance of the phase conductors can be significant. The currents that flow in these components of the circuit impedance constitute reactive power, which transmits no energy to the load. Reactive current flow causes extra losses in the transmission circuit.

The ratio of real power (transmitted to the load) to apparent power is the power factor. As reactive current increases, the reactive power increases and the power factor decreases. For systems with low power factors, losses are higher than for systems with high power factors. Utilities add capacitor banks and other components throughout the system — such as phase-shifting transformers, static VAR compensators, physical transposition of the phase conductors, and flexible AC transmission systems (FACTS) — to control reactive power flow for reduction of losses and stabilization of system voltage.

HVDCN

High voltage direct current (HVDC) is used to transmit large amounts of power over long distances or for interconnections between asynchronous grids. When electrical energy is required to be transmitted over very long distances, it can be more economical to transmit using direct current instead of alternating current.

For a long transmission line, the value of the smaller losses, and reduced construction cost of a DC line, can offset the additional cost of converter stations at each end of the line. Also, at high AC voltages significant amounts of energy are lost due to corona discharge, the capacitance between phases or, in the case of buried cables, between phases and the soil or water in which the cable is buried.

HVDC links are sometimes used to stabilize against control problems with the AC electricity flow. In other words, to transmit AC power as AC when needed in either direction between Seattle and Boston would require the (highly challenging) continuous real-time adjustment of the relative phase of the two electrical grids. With HVDC instead the interconnection would:

- Convert AC in Seattle into HVDC.
- Use HVDC for the three thousand miles of cross country transmission. Then
- Convert the HVDC to locally synchronized AC in Boston, and optionally in other cooperating cities along the transmission route. One prominent example of such a transmission line is the Pacific DC Intertie located in the Western United States.
- *Grid exit:* At the substations, transformers are again used to step the voltage down to a lower voltage for distribution to commercial and residential users. This distribution is accomplished with a combination of sub-transmission (33 kV to 115 kV, varying by country and customer requirements) and distribution (3.3 to 25 kV). Finally, at the point of use, the energy is transformed to low voltage (100 to 600 V, varying by country and customer requirements).
- *Limitations:* The amount of power that can be sent over a transmission line is limited. The origins of the limits vary depending on the length of the line. For a short line, the heating of conductors due to line losses sets a "thermal" limit. If too much current is drawn,

conductors may sag too close to the ground, or conductors and equipment may be damaged by overheating. For intermediate-length lines on the order of 100 km (60 miles), the limit is set by the voltage drop in the line. For longer AC lines, system stability sets the limit to the power that can be transferred. Approximately, the power flowing over an AC line is proportional to the sine of the phase angle between the receiving and transmitting ends. Since this angle varies depending on system loading and generation, it is undesirable for the angle to approach 90 degrees. Very approximately, the allowable product of line length and maximum load is proportional to the square of the system voltage. Series capacitors or phase-shifting transformers are used on long lines to improve stability. High-voltage direct current lines are restricted only by thermal and voltage drop limits, since the phase angle is not material to their operation.

Up to now, it has been almost impossible to foresee the temperature distribution along the cable route, so that the maximum applicable current load was usually set as a compromise between understanding of operation conditions and risk minimization.

The availability of industrial Distributed Temperature Sensing (DTS) systems that measure in real time temperatures all along the cable is a first step in monitoring the transmission system capacity.

This monitoring solution is based on using passive optical fibers as temperature sensors, either integrated directly inside a high voltage cable or mounted externally on the cable insulation. A solution for overhead lines is also available. In this case the optical fibre is integrated into the core of a phase wire of overhead transmission lines (OPPC).

The integrated Dynamic Cable Rating (DCR) or also called Real Time Thermal Rating (RTTR) solution enables not only to continuously monitor the temperature of a high voltage cable circuit in real time, but to safely utilize the existing network capacity to its maximum. Furthermore it provides the ability to the operator to predict the behaviour of the transmission system upon major changes made to its initial operating conditions.

Communications

Operators of long transmission lines require reliable communications for control of the power grid and, often, associated generation and distribution facilities. Fault-sensing protection relays at each end of the line must communicate to monitor the flow of power into and out of the protected line section so that faulted conductors or equipment can be quickly de-energized and the balance of the system restored. Protection of the transmission line from short circuits and other faults is usually so critical that common carrier telecommunications are insufficiently reliable. In remote areas a common carrier may not be available at all. Communication systems associated with a transmission project may use:

- Microwaves
- Power line communication
- Optical fibers

Rarely, and for short distances, a utility will use pilot-wires strung along the transmission line path. Leased circuits from common carriers are not preferred since availability is not under control of the electric power transmission organization.

Transmission lines can also be used to carry data: this is called power-line carrier, or PLC. PLC signals can be easily received with a radio for the long wave range. Optical fibers can be included in the stranded conductors of a transmission line, in the overhead shield wires. These cables are known as *OPGW* or Optical Ground Wire. Sometimes a standalone cable is used, *ADSS* or All Dielectric Self Supporting cable, attached to the transmission line cross arms.

Some jurisdictions, such as Minnesota, prohibit energy transmission companies from selling surplus communication bandwidth or acting as a telecommunications common carrier. Where the regulatory structure permits, the utility can sell capacity in extra "dark fibers" to a common carrier, providing another revenue stream for the line.

Electricity Market Reform

Some regulators, economists, and many power companies regard electric transmission to be a natural monopoly and there are moves in many countries to separately regulate transmission.

Spain was the first country to establish a Regional Transmission Organization. In that country transmission operations and market operations are controlled by separate companies. The transmission system operator is Red Eléctrica de Espana and the wholesale electricity market operator is Operador del Mercado Ibérico de Energía - Polo Español, S.A. (OMEL). Spain's transmission system is interconnected with those of France, Portugal, and Morocco. In the United States and parts of Canada, electrical transmission companies operate independently of generation and distribution companies.

Merchant Transmission

Merchant transmission is an arrangement where a third party constructs and operates electric transmission lines through the franchise area of an unrelated utility. Advocates of merchant transmission claim that this will create competition to construct the most efficient and lowest cost additions to the transmission grid. Merchant transmission projects typically involve DC lines because it is easier to limit flows to paying customers.

The only operating merchant transmission project in the United States is the Cross Sound Cable from Long Island, New York to New Haven, Connecticut, although additional projects have been proposed.There is only one unregulated or market interconnector in Australia: Basslink between Tasmania and Victoria. Two DC links originally implemented as market interconnectors Directlink and Murraylink have been converted to regulated interconnectors. NEMMCO

A major barrier to wider adoption of merchant transmission is the difficulty in identifying who benefits from the facility so that the beneficiaries will pay the toll. Also, it is difficult for a merchant transmission line to compete when the alternative transmission lines are subsidized by other utility businesses.

Health Concerns

Some research has found that exposure to elevated levels of ELF magnetic fields such as those originating from electric power transmission lines may be implicated in a number of adverse health effects. These include, but are not limited to, childhood leukemia, Alzheimer's, adult leukemia, breast cancer, neurodegenerative diseases (such as amyotrophic lateral sclerosis). Miscarriage, and clinical depression. Although there seems to be a small statistical correlation between various diseases and living near power lines, any physical mechanism is not clear. One proposed mechanism is that the electric fields around power lines attract aerosol pollutants.

One possible response to the potential dangers of overhead power lines is to place them underground. According to the British Stakeholder Advisory Group on ELF EMFs, the cost of burying cables at transmission voltages is around GBP 10M/km, compared to GBP 0.5-1M/km for overhead lines. This is mainly due to the limit of the physical properties of the insulation during installation keeping the runs to hundreds of feet between splices, which are most commonly placed in manholes or splice-boxes for repairs.

Underground cables eliminate the electric field and reduce the width over which the magnetic field is elevated. However, in reality, protection from the dangers of electromagnetic (EM) fields is seldom the driving concern when burying power lines. Maintaining underground cables is more expensive; it is usually easier to visually find a problem with an overhead line.

SPECIAL TRANSMISSION

Grids for Railways

In some countries where electric trains run on low

frequency AC (e.g. 16.7 Hz and 25 Hz) power, there are separate single phase traction power networks operated by the railways. These grids are fed by separate generators in some traction powerstations or by traction current converter plants from the public three phase AC network.

Superconducting Cables

High-temperature superconductors promise to revolutionize power distribution by providing lossless transmission of electrical power. The development of superconductors with transition temperatures higher than the boiling point of liquid nitrogen has made the concept of superconducting power lines commercially feasible, at least for high-load applications. It has been estimated that the waste would be halved using this method, since the necessary refrigeration equipment would consume about half the power saved by the elimination of the majority of resistive losses. Such cables are particularly suited to high load density areas such as the business district of large cities, where purchase of a wayleave for cables would be very costly.

Single wire Earth Return

Single wire earth return (SWER) or single wire ground return is a single-wire transmission line for supplying single-phase electrical power for an electrical grid to remote areas at low cost. It is principally used for rural electrification, but also finds use for larger isolated loads such as water pumps, and light rail. Single wire earth return is also used for HVDC over submarine power cables.

Wireless Power Transmission

Every radio transmitter emits power wirelessly. Both Nikola Tesla and Hidetsugu Yagi attempted to devise systems for large scale wireless power transmission. Tesla claimed to have succeeded. Yagi also proposed a similar concept, but the engineering problems proved to be more onerous than conventional systems. His work, however, led to the invention of the Yagi antenna.

Another form of wireless power transmission has been studied for transmission of power from solar power satellites to the earth. A high power array of microwave transmitters would beam power to a rectenna. Major engineering and economic challenges face any solar power satellite project.

Another form is the operation of a crystal radio powered by the radio station it is tuned to, however the energetic efficiency is extremely low. Small scale wireless power was demonstrated as early as 1831 by Michael Faraday and by 1888 Heinrich Rudolf Hertz had proven that natural radio waves exist and can be captured.

Electric Power Systems

Electricity in the large quantities required to supply electric power systems is produced in generating stations, commonly called power plants. Such generating stations, however, should be considered as conversion facilities in which the heat energy of fuel (coal, oil, gas, or uranium) or the hydraulic energy of falling water is converted to electricity. The transmission system carries electric power efficiently and in large amounts from generating stations to consumption areas. Such transmission is also used to interconnect adjacent power systems for mutual assistance in case of emergency and to gain for the interconnected power systems the economies possible in regional operation.

Another approach to high-voltage long-distance transmission is high-voltage direct current (HVDC), which offers the advantages of less costly lines, lower transmission losses, and insensitivity to many system problems that restrict alternating-current systems. Its greatest disadvantage is the need for costly equipment for converting the sending-end power to direct current, and for converting the receiving-end direct-current power to alternating current for distribution to consumers.

As systems grow and the number and size of generating units increase, and as transmission networks expand, higher levels of bulk-power-system reliability are attained through properly coordinated interconnections among separate systems. Most of the electric utilities in the contiguous United States and a large part of Canada now operate as members of power pools, and these pools in turn are interconnected into one gigantic power grid known as the North American Power Systems Interconnection. The operation of this interconnection, in turn, is coordinated by the North American Electric Reliability Council (NERC). Each individual utility in such pools operates independently, but has contractual arrangements with other members in respect to generation additions and scheduling of operation. Their participation in a power pool affords a higher level of service reliability and important economic advantages.

Power delivered by transmission circuits must be stepped down in facilities called substations to voltages more suitable for use in industrial and residential areas.That part of the electric power system that takes power from a bulk-power substation to customers' switches, commonly about 35% of the total plant investment, is called distribution.

The operation and control of the generation-transmission-distribution grid is quite complex because this large system has to operate in synchronism and because many different organizations are responsible for different portions of the grid. In North America and Europe, many public and private electric power companies are interconnected, often across national boundaries. Thus, many organizations have to coordinate to operate the grid, and this coordination can take many forms, from a loose agreement of operational principles to a strong pooling arrangement of operating together.

Power-system operations can be divided into three stages: operations planning, real-time control, and after-the-fact accounting. The main goal is to minimize operations cost while maintaining the reliability (security) of power delivery to customers. Operations planning is the optimal scheduling of generation resources to meet anticipated demand in the next few hours, weeks, or months. This includes the scheduling of water, fossil fuels, and equipment maintenance over many weeks, and the commitment (start-up and shutdown) of generating units over many hours. Real-time control of the system is required to respond to the actual demand of electricity and any unforeseen contingencies (equipment outages). Maintaining security of the system so that a possible contingency cannot disrupt power supply is an integral part of real-time control. After-the-fact accounting is the tracking of purchases and sales of energy between organizations so that billing can be generated.

For loosely coordinated operation of the grid, each utility takes responsibility for the operation of its own portion while exchanging all relevant information. For pool-type operations, a hierarchy is set up where the operational decisions may be made centrally and then implemented by each utility. For a large utility, there may be another level in the hierarchy where the decisions are further distributed to different geographical areas of the same utility. All of this requires significant data communication as well as engineering computation within a utility as well as between utilities. The use of modern computers and communications makes this possible, and the heart of system operations in a utility is the energy control centre.

The monitoring and control of a power system from a centralized control centre became desirable quite early in the development of electric power systems, when generating stations were connected together to supply the same loads. As electrical utilities interconnected and evolved into complex networks of generators, transmission lines, distribution feeders, and loads, the control centre became the operations headquarters for each utility. Since the generation and delivery of electrical energy are controlled from this centre, it is referred to as the energy control centre or energy management system.

Chapter 4

Fossil Fuels-Coal, Oil and Natural Gas

Fossil fuels are produced from animals and plants that were living 300 million years back. Oil, gas, and coal are the common examples of fossil fuels. The term "fossil fuel" is used to describe these fuels because their source dates back to the prehistoric period. These fuels are mostly obtained in the form of deposits under the earth. Fossil fuels have chemical energy stored within them. When these fuels are burnt, the stored energy gets released and serves the purpose. Fossil fuels are used to meet around 85% of the total energy requirement of the world.

ORIGIN OF FOSSIL FUELS

The biogenic theory gives an idea about the origin of fossil fuels. Algae and zooplankton that settled under the water bodies like sea or huge lake in the prehistoric period are the main source of petroleum.

Anoxic conditions are required for the formation of petroleum. On the other hand, terrestrial plants that died in prehistoric period are the main source of coal.

Use of Fossil Fuels

Fossil fuels assume huge importance in the energy scenario of the world. Oil alone meets around 40% of the total requirement of energy. The distillate fuels derived from crude oil are extensively used for transportation, cooking, industrial production and many other purposes. Heating oil, turbo-jet fuel, kerosene, and diesel are some of the examples of distillate fuels. Fossil fuels constitute the chief sources of electrical energy. The thermal power plants make use of coal for power generation. The power plants that use coal contribute 50% of the total demand of energy across the world.

Supply Constraint of Fossil Fuels

Unlike the renewable forms of energy, the sources of fossil fuels are limited. Owing to the increased demand of fossil fuels, world is going to face fuel crisis in future. This is the reason as to why the demand for renewable energy is increasing over the years. Efforts are also on to reduce the use of fossil fuels in order to check the emissions of greenhouse gases.

Types of Fossil Fuels

Fossil fuels are used for a variety of domestic and industrial purposes. These fuels contain high percentages of carbon and generate huge amount of energy when heated. These non-renewable sources of energy are formed as deposits within the inner layers of earth. The remains of animals and vegetations that died millions of years ago formed the reserves of fossil fuels in different parts of the world. Fossil fuels are mostly found in three basic types namely natural gas, coal, and petroleum.

Description of Fossil Fuels

The three types of fossil fuels are obtained from different sources. These fuels are mostly produced from different types of zooplankton and phytoplankton. The transformation process of these ancient organisms took millions of years.

As per the biogenic theory, the remains of certain type of algae or plankton form natural gas after several years of transformation. Methane is the most common form of natural gas found from the inner layers of earth.

Sometimes natural gas is collected from the surface of the petroleum deposits. Owing to its low density, natural gas is normally found over the petroleum reserves. Huge reserves of only natural gas are also found in certain parts of the world.

This form of fossil fuel is normally used for household purposes. It is also more environment-friendly compared to the other forms of fossil fuels.

Coal is formed from the decomposed remains of plants like moss, ferns, and the like through the process of coalification. High atmospheric pressure and temperature are some of the other conditions that are required for the formation of coal. The composition of coal depends on the nature of deposit. More than 1200 different varieties of coal are found in different parts of the world. Graphite is the most carbon-rich form of coal.

Petroleum contains different forms of hydrocarbons both in simple and complex configurations. The remains of oceanic plants and bacteria of prehistoric time formed the deposits of petroleum. It required millions of years for the entire transformation process to complete. Crude petroleum is used to produce a number of distillate fuels like kerosene, turbo-jet fuel and the like. It is also used in plastic production. Petroleum meets nearly 40 percent of the world demand of energy.

Concepts of Fossil Fuels

Fossil fuels are of immense importance for our day-to-day lives. All the three major forms of fossil fuels like oil, natural gas, and coal meet the most of our household and industrial energy requirement. Oil provides mobility to the humanities. Oil is used as fuel in almost all means of transportation like trains, buses, cars, trucks, aircrafts, boats and the like. Coal in extensively used in power generation (electrical power).

The power generated from coal is used to operate different devices. However, the most important use of this power is that it provides light. Natural gas is commonly used for household purposes like heating water, cooking and so on. Sometimes, natural gas is also used to meet the energy requirements of industries.

Compositions of Fossil Fuels

Crude oil is made up of different forms of hydrocarbons with dissimilar molecular weights, chemical properties, and organic structures. Oil collected from natural deposits can be of diverse forms. It may by too viscous or easy-flowing. The different forms of oil are separated and collected from the oil deposits by the procedure of refining. However, crude oil unlike other two forms of fossil fuels can be used only after refining. Natural gas contains hydrocarbons in gaseous forms. Methane is the major component of natural gas. It also contains small amounts of nitrogen, oxygen, hydrogen, propane, ethylene, ethane, and helium. Natural gas is commonly found over oil deposits. It is also found separately within different layers under the earth's surface. Natural gas is drawn by pipes from the deposits and then transported to distant places through a network of pipelines.

Coal is made up of a combination of different chemical compounds like oxygen, carbon, hydrogen, sulfur, nitrogen and some more. Coal is found in a wide variety of forms like bituminous coal, sub-bituminous coal, peat, anthracite, brown coal and so on. Coalification is the process that results in the formation of different varieties of coal. Quality of coal increases as the percentage of carbon rises. The rank of coal improves starting from lignite to low rank coal, high rank coal, and finally to anthracite. Among the different forms of coal, graphite contains the highest percentage of carbon.

There is no denying that fossil fuels constitute the chief source of energy in today's world. However, it produces greenhouse gas, which is harmful for the atmosphere. The sources of fossil fuels are also limited. All these are increasing the demand of alternative renewable sources of energy across the world.

Theory of Fossil Fuels

The fossil fuels are non-renewable sources of energy. These are actually hydrocarbons that remain inside the earth's crust. There are different types of fossil fuels like petroleum, methane, coal and so on. Some of these are made up of pure carbon and others are of low ratio of carbon and hydrogen. According to the most common theory of fossil fuels, it is believed that these fuels were formed from the corpse of animals as well as plants that got buried under the earth's crust due to certain reasons. There are two different theories of fossil fuels that are known as Biogenic theory and Abiogenic theory.

According to the Biogenic theory that is put forward by a number of geologists, the formation of fossil fuels is caused by the heating and compression of the organic products for a long time. As per this theory, oil has been formed from the remains of the algae and zooplankton in the sea bed or lakes. These materials are believed to be of prehistoric age and since then these were under anoxic conditions. On the other hand, coal has been formed from the terrestrial plants. These plants remained buried under the crust of the earth and due to heat and pressure, chemical changes took place.

These changes turned these products into coal. Again, these organic materials, due to heavy pressure and high temperature, changed into kerogen. This process is known as diagenesis. Further rise in the pressure and temperature levels changed kerogen into gaseous or liquid hydrocarbon. This process is termed as catagenesis.

There is another group of geologists who believe in the abiogenic theory. According to this theory, hydrocarbons (of inorganic nature) were present on this planet since the very beginning. Now, the densities of the petroleum hydrocarbons are lower than the aqueous pore fluids and because of this, these hydrocarbons move upward and get trapped in the oil reservoirs. This movement was made possible by the fracture networks. This theory further demonstrates that the biomarkers that remain in the petroleum are actually due to the microbial life-forms that remain in the rocks. Although this is an important theory of fossil fuels but there are not enough evidences to support this theory. Both these theories hold that millions of years are needed for the creation of fossil fuels. The reservoirs of these fuels are becoming exhausted very rapidly and the rate of new fossil fuel formation is very low. Thus, these fuels are marked as non-renewable fuels.

Energy Cost of Fossil Fuels

The term fossil fuel means the hydrocarbons found in the

top layer of Earth's crust. Combustible oil, natural gas, and coal are prime examples of fossil fuel. The energy costs of fossil fuels could be classified into 2 broad categories: One that we pay money for, and the other, we do not. The purchase price of any fossil fuel like petrol includes the drilling cost, labor working in refining plants, and the transport of the fuel to retail points. The hidden costs, that we do not pay for, include global warming and associated health problems.

The environmental fallout of fossil fuel use includes the many tangible negative influences that can be felt. The rapid deterioration of air quality is being felt throughout the world. Acid rain is real and happening at regular intervals.

The specter of global warming is threatening the coastal territories of countries all over the globe. The hidden costs of fossil fuels are:

- *Water and land pollution:* Transportation of liquid fossil fuels like petrol and diesel involve a high risk of leak and spillage. Gasoline (petrol) spills in water severely damages the immediate aquatic environment like rivers and seas. Animal and plant life are negatively affected by oil spills. The mining of solid fossil fuels like coal also contributes to significant water pollution. The pyrite in coal is washed away during processing to pollute rivers and streams.
- *Air pollution:* Burning fossil fuel generates nitrogen oxides, carbon monoxide, and sulfur oxides. Hydrocarbons are also disseminated into the air. The particulates that are suspended in air contribute to air pollution. The hydrocarbons and nitrogen oxides in the air combine to form smog. Transportation vehicles like cars and buses contribute the most to air or atmospheric pollution. Nitrogen oxides and sulfur oxides are spewed by cars. Acid rain occurs when these two combine with the water vapour present in clouds.
- *Global warming:* Carbon monoxide is the result of burning fossil fuels. The temperature of the earth has increased by approximately 1° centigrade since the

1800s. This has led to glacier melting in the Polar Regions, thus increasing the water content in the seas. This has further led to inundation of river deltas, wetlands, and coastal areas. Extreme weather is reported throughout the world.

- *Pollution by heat:* The production of electricity by burning fossil fuels involve the generation of heat as a collateral product. The heat is released into the atmosphere. Water that is employed as a coolant also gets heated. This heated water, when returned, may disturb the aquatic ecosystem.

Utilization of Fossil Fuels

One of the major forms of fossil fuels is coal. There are several forms of coal like anthracite coal, peat, bituminous and lignite. Peat and lignite are not high on productivity as they give off only negligible amount of energy for every unit of fuel. Lignite is still more useful compared to peat as it may be vaporized to obtain a clean burning fuel. At times lignite is also liquified in order to obtain fractions of liquid petroleum. However these processes are expensive. Anthracite and bituminous coal are processed after being extracted from the mine and then can be used as fuels.

One of the major issues regarding the utilization of coal is that they are very expensive to extract. Besides the various types of structures that need to be constructed and the sophisticated machinery that need to be used for these purposes there are a lot of dangers like dilapidation of the structures as well as explosion. The laborers who work in these mines may often suffer from respiratory diseases as a result of breathing the poisonous air present in their working environment. The transportation of coals is also a major issue as a result of the costs involved.

Crude or raw petroleum is yet another fossil fuel as well. It may be described as a combination of several elements like dissolved gases, solids and liquids. Crude petroleum is considered to be more advantageous compared to coal. It is easier to extract raw petrol as it is liquid and thus can be drawn to the surface by putting pressure. It is also cost effective compared to coal. However there are certain disadvantages of raw petroleum as well as it can be transferred by ships or pipelines. This is also less time consuming compared to the transportation of coal, which has to be done by railroads or ships that are headed towards the oceans.

REDUCING DEPENDENCY ON FOSSIL FUELS

Importance of Sustainable Energy

Reducing dependency on fossil fuels is a major challenge for most economically advanced countries of the world as there is a very important link between the usage of fossil fuels and the economic conditions of individual. One of the major reasons behind the increasing demand to reduce the reliance on fossil fuels is the fact that they are expected to run out at a certain point of time in the future. This is where sustainable energy is expected to be useful as it would last longer than any other form of fuel source.

Reducing Dependency on Fossil Fuels in USA

Countries like the United States of America have been focusing long and hard on ways to find out alternatives to the fossil fuels. One of the major areas where the Dependency on fossil fuel needs to be reduced is electricity. A lot of countries use fossil fuels like coal, natural gases and oil for the purpose of generating electricity. With the alternative sources of generating electricity it would be possible for countries like these to import lesser oil, natural gases and coal and thereby not exhaust the limited resources. However, petroleum and natural gases are also used in other sectors of the economy like transportation and several industries for example. Decrease in the usage of fossil fuels would be beneficial from an ecological point of view as well since less use fossil fuels mean lesser emissions of harmful gases.

Ways of Reducing Dependency on Fossil Fuels

There are certain ways in which the Dependency on the

fossil fuels may be reduced. One major way is to change or replace the technologies that are used in the construction of the automobiles as they are the major sources of air pollution. Experts are of the opinion that people need to use electric batteries in their cars. The principal reason behind this argument is that this using battery driven cars would reduce the Dependency on the fossil fuels.

Experts Opinion on Reducing Dependency on Fossil Fuels

Experts from major oil companies have been instructing people to use alternative fuels. They have suggested a number of steps that may be followed in order to reduce the dependency on fossil fuels. Those steps may be suggested as below:

- Using diesel in the cars
- Turning lawns into organic gardens
- Using solar energy at home
- Using wood for the purposes of heating
- Using telecommunications in the professional life

People connected to the fields of science and technologies have been coming up with new gases. In the United States of America the scientists have been busy in creating new gases like syngas, which is a combination of hydrogen and carbon oxides. These are supposed to replace gases like ethanol, which is supposed to have lesser environmental value. Biomass would be important in this context.

Advantages of Reducing Dependency on Fossil Fuels

Reducing dependency on fossil fuels has its economic advantages as well. It has been calculated that the engines operated with electricity are far more efficient in terms of mileage than the vehicles that run on fossil fuels like petroleum for example. It has been assumed that any possible increase in the generation of electricity is supposed to be a combination of several types of alternative energy resources. Those may be mentioned as below:

- Hydroelectricity
- Nuclear energy
- Concentrated solar thermal energy
- Wind power

- Geothermal energy
- Solar cells

COMMERCIAL USES

Once fossil fuel has been extracted and processed, it can be burned for direct uses, such as to power cars or heat homes, or it can be combusted for the generation of electrical power.

Direct Combustion

Fossil fuels are primarily burned to produce energy. This energy is used to power automobiles, trucks, airplanes, trains, and ships around the world; to fuel industrial manufacturing processes; and to provide heat, light, air conditioning, and energy for homes and businesses. About two-fifths of all energy consumed in the United States is used by industry, one-third by homes and businesses, and about one-fourth by transportation.

To provide fuel for transportation, petroleum is refined into gasoline, diesel fuel, jet fuel, and other derivatives used in most of the world's automobiles, trucks, trains, aircraft, and ships. In the United States, transportation accounts for about two-thirds of total petroleum consumption—more than two-thirds of which is burned as automobile gasoline.

Demand for natural gas, historically considered a waste by-product of petroleum and coal mining, is growing in business and industry because it is a cleaner-burning fuel than petroleum or coal. Natural gas, which can be piped directly to commercial plants or individual residences and used on demand, is used for heating and for air conditioning. Residential uses of natural gas also include fuel for stoves and other heating appliances.

Electricity Generation

In addition to direct combustion for commercial uses, fossil fuels are also burned to generate most of the world's electric power. In 2003 fossil fuel–fired power plants produced 65 percent of the world's electrical power, down from 71 percent in the late 1970s. In 2003 the world's remaining electricity supply was generated primarily by hydroelectric power (17 percent) and nuclear fission (16 percent), with solar, geothermal, and other sources accounting for a relatively small amount.

FORMATION OF FOSSIL FUELS

Fossil fuels formed from ancient organisms that died and were buried under layers of accumulating sediment. As additional sediment layers built up over these organic deposits, the material was subjected to increasing temperatures and pressures. Over millions of years, these physical conditions chemically transformed the organic material into hydrocarbons. Most organic debris is destroyed at the earth's surface by oxidation or by consumption by microorganisms.

Organic material that survives to become buried under sediments or deposited in other oxygen-poor environments begins a series of chemical and biological transformations that may ultimately result in petroleum, natural gas, or coal. Many such deposits occur in sedimentary basins, and along continental shelves. Sediments may accumulate to depths of several thousand feet in a basin, exerting pressures up to one hundred million pascals (tens of thousands of pounds per square inch) and temperatures of several hundred degrees on the organic material. Over millions of years, these conditions can chemically transform the organic material into petroleum, natural gas, coal, or other types of fossil fuels.

Petroleum Formation

Petroleum formed chiefly from ancient, microscopic plants and bacteria that lived in the ocean and saltwater seas. When these microorganisms died and settled to the seafloor, they mixed with sand and silt to form organic-rich mud. As layers of sediment accumulated over this organic ooze, the mud was gradually heated and slowly compressed into shale or mudstone, chemically transforming the organic material into petroleum and natural gas. Sometimes, the petroleum and natural gas would slowly fill the tiny holes within nearby porous rocks, which geologists call reservoir rocks. Because these porous rocks were usually filled with water, the liquid and gaseous hydrocarbons (which are less dense and lighter than water) migrated upward, through the earth's crust, sometimes for long distances.

A portion of these hydrocarbons would eventually encounter an impermeable (nonporous) layer of rock in an anticline, salt dome, fault trap, or stratigraphic trap. The impermeable rock would trap the hydrocarbons, creating a reservoir of petroleum and natural gas. Exploration geologists seek these underground formations because they often contain recoverable petroleum deposits. The fluids and gases caught in these geologic traps typically separate into three layers: water (highest density, bottom layer), petroleum (middle layer), and natural gas (low density, top layer).

Coal Formation

Coal is a solid fossil fuel formed from ancient plants—including trees, ferns, and mosses—that grew in swamps and bogs or along coastal shorelines. Generations of these plants died and were gradually buried under layers of sediment. As the sedimentary overburden increased, the organic material was subjected to increasing heat and pressure that caused the organic material to undergo a number of transitional states to form coal.

The mounting pressure and temperature caused the original organic material, which was rich in carbon, hydrogen, and oxygen, to become increasingly carbon-rich and hydrogen- and oxygen-poor. The successive stages of coal formation are peat (partially carbonized plant matter), lignite (soft brownish-black coal with low carbon content), subbituminous coal (soft coal with intermediate carbon content), bituminous coal (soft coal with higher carbon and lower moisture content than subbituminous coal), and anthracite (hard coal with highest carbon content and lowest moisture content). Because anthracite is the most carbon-rich, moisture-deficient form of coal, it has the highest heating value.

Natural Gas Formation

Most natural gas is formed from plankton—tiny water-dwelling organisms, including algae and protozoans—that accumulated on the ocean floor as they died. These organisms were slowly buried and compressed under layers of sediment. Over millions of years, the pressure and heat generated by overlying sediments converted this organic material into natural gas.

Natural gas is composed primarily of methane and other light hydrocarbons. Natural gas frequently migrates through porous and fractured reservoir rock with petroleum and subsequently accumulates in underground reservoirs. Because of its light density relative to petroleum, natural gas forms a layer over the petroleum. Natural gas may also form in coal deposits, where it is often found dispersed throughout the pores and fractures of the coal bed.

Other Fossil Fuels

Geologists have identified immense deposits of other hydrocarbons, including gas hydrates (methane and water), tar sands, and oil shale. Vast deposits of gas hydrates are contained in ocean sediments and in shallow polar soils. In these marine and polar environments, methane molecules are encased in a crystalline structure with water molecules.

This crystalline solid is known as gas hydrate. Because technology for the commercial extraction of gas hydrates has not yet been developed, this type of fossil fuel is not included in most world energy resource estimates. Energy completed the drilling of a well to retrieve core samples of gas hydrates found in the Prudhoe Bay region of Alaska's North Slope. By identifying the nature of these gas hydrates, the Energy Department hoped to evaluate the potential of natural gas production from this region.

Tar sands are heavy, asphaltlike hydrocarbons found in sandstone. Tar sands form where petroleum migrates upward into deposits of sand or consolidated sandstone. When the petroleum is exposed to water and bacteria present in the sandstone, the hydrocarbons often degrade over time into heavier, asphaltlike bitumen. Oil shale is a fine-grained rock containing high concentrations of a waxy organic material known as *kerogen*. Oil shale forms on lake and ocean bottoms where dead algae, spores, and other microorganisms died millions of years ago and accumulated in mud and silt.

The increasing pressure and temperature from the buildup of overlying sediments transformed the organic material into kerogen and compacted the mud and silt into oil shale. However, this pressure and heat was insufficient to chemically break down the kerogen into petroleum. Because the hydrocarbons contained in tar sand and oil shale are not fluids, these hydrocarbons are more difficult and costly to recover than liquid petroleum.

WORLD FOSSIL FUEL SUPPLY

Because the global economy is powered by fossil fuels, it is critical to know how long world reserves will last. However, estimating the world's remaining fossil fuel reserves requires extensive information, including comprehensive geological maps of the world's sedimentary basins, models of energy production systems, and data showing world energy consumption patterns and trends.

Reserves and Resources

When estimating the world's fossil fuel supply, experts distinguish between *reserves* and *resources*. Reserves are fossil fuel deposits that have already been discovered and are immediately available. Resources are fossil fuel deposits that geologists believe are located in certain sedimentary basins, but have not yet been discovered. Because geologists base fossil fuel resource estimates on the location, extent, and formation of deposits recovered in geologically similar basins, resource estimates are less certain than reserve estimates. Both reserve and resource estimates are revised as data about new and existing deposits become available.

Fossil fuel reserves can be divided into *proved reserves* and *inferred reserves*. Proved reserves are deposits that have been measured, sampled, and evaluated for production quality. Inferred reserves have been discovered but have not been measured or evaluated. The definition of fossil fuel resources can be narrowed to *technically recoverable resources*. This definition does not consider whether a deposit can be extracted economically, but only whether the fossil fuel can be recovered using existing technology. By definition, the world fossil fuel supply increases as technological advances are made allowing previously unrecoverable resources to be extracted and processed.

World Energy Data

Worldwide deposits of fossil fuels are finite. Some experts use data on world energy deposits to estimate how many years world energy supplies will last at current and projected consumption rates.

At the beginning of the 21st century, the world reserves of petroleum were estimated to be roughly 1.3 trillion barrels. By 2003 worldwide consumption of petroleum totaled 29 billion barrels a year.

The world's natural gas reserves were estimated to be roughly 170 trillion cubic meters (6,000 trillion cubic feet). Worldwide consumption of natural gas by 2003 totaled 2.7 trillion cubic meters (96 trillion cubic feet) a year. At the beginning of the 21st century the world coal reserves were estimated to be roughly 1 trillion metric tons, and by 2003 worldwide consumption of coal totaled 5 billion metric tons a year. Total worldwide energy consumption is expected to grow at about 2.2 percent per year until 2015.

Theoretical models can be developed to estimate how many years the world fossil fuel supply will last. However, these models are complicated by technological advances in the energy production industry, unexpected discoveries of new fossil fuel deposits, and political, social, and economic factors that influence energy production and consumption.

Because fossil fuels are being consumed at much faster rates than they are produced in the earth's crust, humankind will eventually deplete these nonrenewable resources. While it is unclear how far in the future this will happen, there is evidence that some regions are becoming depleted in certain types of fossil fuels. Production of crude petroleum in the

United States peaked in 1970. Today the United States imports significantly higher proportions of its petroleum needs.

Alternative Energy Sources

The prospect of reducing the world's dependence on fossil fuels is problematic. Alternative energy industries, such as nuclear energy, hydroelectric energy, solar energy, wind energy, and geothermal energy exist, but these energy sources currently only account for a combined 14 percent of energy consumed worldwide. To date, alternative energy sources have been hindered by technological and environmental difficulties. For instance, although the uranium that fuels nuclear power is abundant, the risk of nuclear accidents and the difficulty associated with safe disposal of radioactive waste have led to the decline of the nuclear power industry. Conversely, solar and wind power seem environmentally safe, but they are unreliable as steady sources of energy. As global energy consumption grows each year, development of certain alternative energy sources becomes increasingly important.

ENVIRONMENTAL EFFECTS OF USING FOSSIL FUELS

Acid rain and global warming are two of the most serious environmental issues related to large-scale fossil fuel combustion. Other environmental problems, such as land reclamation and oil spills, are also associated with the mining and transporting of fossil fuels.

Acid Rain

When fossil fuels are burned, sulfur, nitrogen, and carbon combine with oxygen to form compounds known as *oxides*. When these oxides are released into the air, they react chemically with atmospheric water vapour, forming sulfuric acid, nitric acid, and carbonic acid, respectively. These acid-containing water vapors—commonly known as acid rain—enter the water cycle and can subsequently harm the biological quality of forests, soils, lakes, and streams. The U.S. Clean Air Act has significantly reduced emissions of acid-producing oxides. The legally mandated removal of sulfur-bearing compounds from coal prior to burning has significantly reduced atmospheric levels of sulfur dioxide. Environmental laws also require use of pollution-trapping equipment, such as air scrubbers. Scrubbers—devices installed inside the smokestacks at a coal-burning plant—filter out sulfur-dioxide vapors and other compounds before these pollutants enter the atmosphere.

Ash Particles

Combustion of fossil fuels produces unburned fuel particles, known as ash. In the past, coal-fired power plants have emitted large amounts of ash into the atmosphere. However, government regulations also require that emissions containing ash be scrubbed or that particles otherwise be trapped to reduce this source of air pollution. While petroleum and natural gas generate less ash than coal, air pollution from fuel ash produced by automobiles may be a problem in cities where diesel and gasoline vehicles are concentrated.

Global Warming

Carbon dioxide is a major by-product of fossil fuel combustion, and it is what scientists call a greenhouse gas. Greenhouse gases absorb solar heat radiated from the earth's surface and retain this heat, keeping the earth warm and habitable for living organisms. Rapid industrialization through the 19th and 20th centuries, however, has resulted in increasing fossil fuel emissions, raising the percentage of carbon dioxide in the atmosphere by about 28 percent. This dramatic increase in carbon dioxide has led some scientists to predict a global warming scenario that could cause numerous environmental problems, including disrupted weather patterns and polar ice cap melting.

Although it is extremely difficult to attribute observed global temperature changes directly to fossil fuel combustion, some countries are working together to lower emissions of carbon dioxide from fossil fuels. One proposal is to establish a system requiring companies to pay to emit carbon dioxide a specified level. This payment could take several forms, including: (1) purchasing the rights to pollute from a company whose carbon dioxide emissions fall below the specified level; (2) purchasing and then preserving forests, which absorb carbon dioxide; and (3) paying to upgrade a carbon dioxide emitting plant in a lesser-developed country, lowering the upgraded plant's carbon dioxide emissions.

Petroleum Recovery and Transportation

Environmental problems are created by drilling oil wells and extracting fluids because the petroleum pumped up from deep reservoir rocks is often accompanied by large volumes of salt water. This brine contains numerous impurities, so it must either be injected back into the reservoir rocks or treated for safe surface disposal.

Petroleum usually must also be transported long distances by tanker or pipeline to reach a refinery. Transport of petroleum occasionally leads to accidental spills. Oil spills, especially in large volumes, can be detrimental to wildlife and habitat.

Coal Mining

Surface coal mining operations, often called strip mines, use massive shovels to remove soil and rock overlying the coal, disrupting the natural landscape. However, new land reclamation methods, driven by stringent laws and regulations, now require mining companies to restore strip-mined landscapes to nearly premined conditions.

Another environmental problem associated with coal mining occurs when freshly excavated coal beds are exposed to air. Sulfur-bearing compounds in the coal oxidize in the presence of water to form sulfuric acid. When this sulfuric acid solution, known as *acid mine drainage,* enters surface water and groundwater, it can be detrimental to water quality and aquatic life. Efforts are currently underway to remove sulfuric acid from mine drainage before it reaches rivers, lakes, and streams. Scientists are studying whether artificial wetlands have the ability to neutralize acid mine drainage.

COAL

Coal, a combustible organic rock composed primarily of carbon, hydrogen, and oxygen. Coal is burned to produce energy and is used to manufacture steel. It is also an important source of chemicals used to make medicine, fertilizers, pesticides, and other products. Coal comes from ancient plants buried over millions of years in Earth's crust, its outermost layer. Coal, petroleum, natural gas, and oil shale are all known as fossil fuels because they come from the remains of ancient life buried deep in the crust.

Coal is rich in hydrocarbons (compounds made up of the elements hydrogen and carbon). All life forms contain hydrocarbons, and in general, material that contains hydrocarbons is called organic material. Coal originally formed from ancient plants that died, decomposed, and were buried under layers of sediment during the Carboniferous Period, about 360 million to 290 million years ago. As more and more layers of sediment formed over this decomposed plant material, the overburden exerted increasing heat and weight on the organic matter. Over millions of years, these physical conditions caused coal to form from the carbon, hydrogen, oxygen, nitrogen, sulfur, and inorganic mineral compounds in the plant matter. The coal formed in layers known as seams.

Plant matter changes into coal in stages. In each successive stage, higher pressure and heat from the accumulating overburden increase the carbon content of the plant matter and drive out more of its moisture content. Scientists classify coal according to its fixed carbon content, or the amount of carbon the coal produces when heated under controlled conditions. Higher grades of coal have a higher fixed carbon content.

Modern Uses of Coal

Eighty-six percent of the coal used in the United States is burned by electric power plants to produce electricity. When burned, coal generates energy in the form of heat. In a power plant that uses coal as fuel, this heat converts water into steam, which is pressurized to spin the shaft of a turbine. This spinning shaft drives a generator that converts the mechanical energy of the rotation into electric power.

Coal is also used in the steel industry. The steel industry uses coal by first heating it and converting it into coke, a hard substance consisting of nearly pure carbon. The coke is combined with iron ore and limestone. Then the mixture is heated to produce iron. Other industries use different coal gases given off during the coke-forming process to make fertilizers, solvents, medicine, pesticides, and other products.

Fuel companies convert coal into easily transportable gas or liquid fuels. Coal-based vapour fuels are produced through the process of *gasification*. Gasification may be accomplished either at the site of the coalmine or in processing plants. In processing plants, the coal is heated in the presence of steam and oxygen to produce *synthesis gas*, a mixture of carbon monoxide, hydrogen, and methane used directly as fuel or refined into cleaner-burning gas.

On-site gasification is accomplished by controlled, incomplete burning of an underground coal bed while adding air and steam. To do this, workers ignite the coal bed, pump air and steam underground into the burning coal, and then pump the resulting gases from the ground. Once the gases are withdrawn, they may be burned to produce heat or generate electricity. Or they may be used in synthetic gases to produce chemicals or to help create liquid fuels.

Liquefaction processes convert coal into a liquid fuel that has a composition similar to that of crude petroleum. Coal can be liquefied either by direct or indirect processes. However, because coal is a hydrogen-deficient hydrocarbon, any process used to convert coal to liquid or other alternative fuels must add hydrogen. Four general methods are used for liquefaction:

- Pyrolysis and hydrocarbonization, in which coal is heated in the absence of air or in a stream of hydrogen;
- Solvent extraction, in which coal hydrocarbons are selectively dissolved and hydrogen is added to produce the desired liquids;
- Catalytic liquefaction, in which hydrogenation takes place in the presence of a catalyst;

- Indirect liquefaction, in which carbon monoxide and hydrogen are combined in the presence of a catalyst.

Coal Formation

Coal is a sedimentary rock formed from plants that flourished millions of years ago when tropical swamps covered large areas of the world. Lush vegetation, such as early club mosses, horsetails, and enormous ferns, thrived in these swamps. Generations of this vegetation died and settled to the swamp bottom, and over time the organic material lost oxygen and hydrogen, leaving the material with a high percentage of carbon. Layers of mud and sand accumulated over the decomposed plant matter, compressing and hardening the organic material as the sediments deepened. Over millions of years, deepening sediment layers, known as overburden, exerted tremendous heat and pressure on the underlying plant matter, which eventually became coal.

Before decayed plant material forms coal, the plant material forms a dark brown, compact organic material known as peat. Although peat will burn when dried, it has a low carbon and high moisture content relative to coal. Most of coal's heating value comes from carbon, whereas inorganic materials, such as moisture and minerals, detract from its heating value.

For this reason, peat is a less efficient fuel source than coal. Over time, as layers of sediment accumulate over the peat, this organic material forms lignite, the lowest grade of coal. As the thickening geologic overburden gradually drives moisture from the coal and increases its fixed carbon content, coal evolves from lignite into successively higher-graded coals: subbituminous coal, bituminous coal, and anthracite. Anthracite, the highest rank of coal, has nearly twice the heating value of lignite.

Coal formation began during the Carboniferous Period (known as the *first coal age*), which spanned 360 million to 290 million years ago. Coal formation continued throughout the Permian, Triassic, Jurassic, Cretaceous, and Tertiary Periods (known collectively as the *second coal age*), which spanned 290 million to 1.6 million years ago. Coals formed during the first coal age are older, so they are generally located deeper in Earth's crust. The greater heat and pressures at these depths produce higher-grade coals such as anthracite and bituminous coals. Conversely, coals formed during the second coal age under less intense heat and pressure are generally located at shallower depths. Consequently, these coals tend to be lower-grade subbituminous and lignite coals.

COMPONENTS OF COAL

Coal contains organic (carbon-containing) compounds transformed from ancient plant material. The original plant material was composed of cellulose, the reinforcing material in plant cell walls; lignin, the substance that cements plant cells together; tannins, a class of compounds in leaves and stems; and other organic compounds, such as fats and waxes.

In addition to carbon, these organic compounds contain hydrogen, oxygen, nitrogen, and sulfur. After a plant dies and begins to decay on a swamp bottom, hydrogen and oxygen (and smaller amounts of other elements) gradually dissociate from the plant matter, increasing its relative carbon content.

Coal also contains inorganic components, known as *ash*. Ash includes minerals such as pyrite and marcasite formed from metals that accumulated in the living tissues of the ancient plants. Quartz, clay, and other minerals are also added to coal deposits by wind and groundwater. Ash lowers the fixed carbon content of coal, decreasing its heating value.

COAL DEPOSITS AND RESERVES

Although coal deposits exist in nearly every region of the world, commercially significant coal resources occur only in Europe, Asia, Australia, and North America. Commercially significant coal deposits occur in sedimentary rock basins, typically sandwiched as layers called beds or seams between layers of sandstone and shale.

When experts develop estimates of the world's coal supply, they distinguish between coal *reserves* and *resources*. Reserves are coal deposits that can be mined profitably with existing technology—that is, with current equipment and methods. Resources are an estimate of the world's total coal deposits, regardless of whether the deposits are commercially accessible. Exploration geologists have found and mapped the world's most extensive coal beds. At the beginning of 2001, global coal reserves were estimated at 984.2 billion metric tons, in which 1 metric ton equals 1,016 kg. These reserves occurred in the following regions by order of importance: the Asia Pacific, including Australia, 29.7 percent; North

Coal deposits in the United Kingdom, which led the world in coal production until the 20th century, extend throughout parts of England, Wales, and southern Scotland.

Coalfields in western Europe underlie the Saar and Ruhr valleys in Germany, the Alsace region of France, and areas of Belgium. Coalfields in central Europe extend throughout parts of Poland, the Czech Republic, and Hungary. The most extensive and valuable coalfield in eastern Europe is the Donets Basin, between the Dnieper and Don rivers. Large coal deposits in Russia are being mined in the Kuznetsk Basin in southern Siberia. Coalfields underlying northwestern China are among the largest in the world. Mining of these fields began in the 20th century.

COAL MINING

Coal mining is the removal of coal from the ground. The mining method employed to extract the coal depends on the following criteria:

- Seam thickness,
- The overburden thickness,
- The ease of removal of the overburden,
- The ease with which a shaft can be sunk to reach the coal seam,
- The amount of coal extracted relative to the amount that cannot be removed, and
- The market demand for the coal.

The two types of mining methods are surface mining and underground mining. In surface mining the layers of rock or soil overlying a coal seam are first removed after which the coal is extracted from the exposed seam. In underground mining, a shaft is dug to reach the coal seam. Currently, underground mining accounts for approximately 60 percent of the world recovery of coal.

Surface Mining

Surface mining is used to reach coal reserves that are too shallow to be reached by other mining methods. Types of surface mining include open-pit mining, drift mining, slope mining, contour mining, and auger mining.

Open-pit Mining

In open-pit mining, or strip mining, earth-moving equipment is used to remove the rocky overburden and then huge mechanical shovels scoop coal up from the underlying deposit. The modern coal industry has developed some of the largest industrial equipment ever made, including shovels capable of holding 290 metric tons of coal.

To reach the coal, bulldozers clear the vegetation and soil. Depending on the hardness and depth of the exposed sedimentary rocks, these rocky layers may be shattered with explosives. To do this, workers drill *blast holes* into the overlying sedimentary rock, fill these holes with explosives, and then blast the overburden to fracture the rock. Once the broken rock is removed, coal is shoveled from the underlying deposit into giant earth-moving trucks for transport.

Drift Mining

Drift mining is used when a horizontal seam of coal emerges at the surface on the side of a hill or mountain, and the opening into the mine can be made directly into the coal seam. This type of mining is generally the easiest and most economical type because excavation through rock is not necessary. If coal is available in this manner, it is likely to be mined.

Underground Mining

Underground, or deep, mining occurs when coal is

extracted from a seam without removal of the overlying strata. Miners build a shaft mine that enters the earth through a vertical opening and descends from the surface to the coal seam. In the mine, the coal is extracted from the seam by various methods, including conventional mining, continuous mining, longwall mining, and room-and-pillar mining.

Conventional Mining

Conventional mining, also called cyclic mining, involves a sequence of operations that proceed in the following order:

- Supporting the roof,
- Ventilation,
- Cutting,
- Drilling,
- Blasting,
- Coal removal, and
- Loading. First, miners make the roof above the seam safe and stable by timbering or by roof bolting, processes intended to prevent the roof from collapsing. At the same time, they create ventilation openings so that dangerous gases can escape and fresh air can reach the miners. Then one or more slots—a few centimeters wide and extending for several meters into the coal—are cut along the face of the coal seam, also known as the wall face, by a large, mobile cutting machine. The cut, or slot, provides easy access to the face and facilitates the breaking up of the coal, which is usually blasted from the seam by explosives known as permissible explosives. This type of explosive produces an almost flame-free explosion and markedly reduces the amount of noxious fumes in comparison with conventional explosives. The coal may then be transported by rubber-tired electric vehicles (shuttle cars) or by chain (or belt) conveyor systems.

Continuous Mining

Continuous mining involves the use of a single machine known as a continuous miner that breaks the coal mechanically and loads it for transport. This mobile machine has a series of metal-studded rotating drums that gouge coal from the face of the coal seam. One continuous miner can mechanically break apart about 1.8 metric tons of coal per hour. Roof support is then installed, ventilation is advanced, and the coalface is ready for the next cycle. The method used to transport the coal requires the installation of mobile belt conveyors.

Longwall Mining

The longwall mining system uses a remote-controlled self-advancing roof in which large blocks of coal are completely extracted in a continuous operation. Hydraulic or self-advancing jacks, known as chocks, support the roof at the immediate face as the coal is removed. As the face advances, the roof is allowed to collapse behind the remote-controlled, roof-building machinery. Miners then remove the fallen coal. Coal recovery is comparable to that attainable with the conventional or continuous mining systems.

Room-and-Pillar Mining

Room-and-pillar mining is a means of developing a coalface and, at the same time, retaining supports for the roof. With this technique, rooms are developed from large, parallel tunnels driven into the solid coal, and the intervening pillars of coal are used to support the roof. The percentage of coal recovered from a seam depends on the number and size of protective pillars of coal thought necessary to support the roof safely. Workers may remove some coal pillars just before closing the mine.

COAL MINING SAFETY

Coalmines are hazardous operations. In the 20th century, more than 100,000 miners died working in coalmines. Many accidents were caused by the mine structure failing through roof collapse or *rock bursts* (coal pillars exploding from the weight of excessive overburden). Other dangers that miners face include toxic or explosive gases released as the coal is mined, dangerous coal dust, and fires. Fires result when flammable gases trapped in the coal, such as methane, are released during mining operations and accidentally ignited.

Provision of adequate ventilation is an essential safety feature of underground coal mining. Not all of this ventilation is required to enable miners to work in comfort. Most of it is required to dilute the harmful gases, frequently termed damps, produced during mining operations.

Mine Failure

In room-and-pillar mines deep underground, the extreme weight of the overlying rock can break the pillars down, either gradually or in a violent collapse. If the mine roof or floor consists of softer material such as clay, the massive weight of the overburden can slowly push the pillars into the floor or ceiling, endangering the stability of the mine. When the mine roof and floor consist of particularly hard rock, massive overhead weight can overload the pillars, sometimes causing a spontaneous collapse or rock burst.

GASES

Because coal and natural gas form by similar natural processes, methane (a principal component of natural gas) is often trapped inside coal deposits. Gases trapped in deeper coal beds have a harder time escaping. Consequently, high-grade coals, which are typically buried deeper than low-grade coals, often contain more methane in the pores and fractures of the deposit. As coalminers saw or blast into a coal deposit, they can release these methane pockets, which may explode spontaneously, often with deadly results. Miners use a technique called *methane drainage* to reduce dangerous releases of methane. Before mining machinery cuts into the wall face, holes are drilled into the coal and methane is drawn out and piped to the surface.

Coalminers also risk being exposed to other deadly gases, including carbon monoxide, a poisonous by-product of partially burned coal. Carbon monoxide is deadly in quantities as little as 1 percent. It is especially prevalent in underground mines after a methane explosion. In the early 1800s, after a gas explosion, coalminers used canaries to test for carbon monoxide. If the canary died, the miners increased ventilation in the mine to remove the carbon monoxide. The miners then conducted the same test with another canary and repeated the process until a bird survived. Miners also tested for carbon monoxide and methane with a small flame. If the flame's size increased, methane was present in the air; if the flame went out, carbon monoxide was present.

Other dangerous gases locked inside coal deposits include hydrogen sulfide, a poisonous, colorless gas with an odor of rotten eggs, and carbon dioxide, a colorless, odorless gas. To prevent injury from inhaling these gases, underground coalmines must be sufficiently ventilated. To do this fresh air is continuously pumped through ventilation holes to push out or dilute dangerous gases and provide air for miners to breathe.

Coal Dust

As coal is blasted, shredded, and hauled in a mine, large amounts of coal and silica dust are produced. Coal dust is extremely flammable, and if ignited it can be more violently explosive than methane. Miners can reduce the buildup of coal dust by injecting pressurized water into coal beds before the coal is blasted or cut, by spraying water at all points where dust is likely to be formed, and by installing dust extraction units at strategic points.

Miners who inhale coal dust over a prolonged period can damage their lung tissue. Often, these miners develop spots, lumps, or fibrous growths in their lungs, a condition known as black lung disease, a type of pneumoconiosis. Black lung disease can develop into other, often fatal illnesses, including heart disease, emphysema, and cancer. To protect miners from black lung, many mines are equipped with coal-dust filtering units. Miners who operate drilling, cutting, or loading machinery should wear masks at all times.

Fires

Coalmine fires can be triggered during routine mining operations. Sparks generated by mining equipment can ignite explosive gases, coal dust, and even the coal bed itself. Because coal beds provide an almost inexhaustible fuel source, once a coal seam is ignited, it can be extremely difficult to extinguish. The intense heat generated by burning coal can rupture the overlying rock strata, sometimes causing the roof to collapse. Uncontrollable fires in some coal deposits have continued burning for years, posing a danger to local communities.

COAL PREPARATION

As-mined coal, also known as run-of-mine coal, often contains unwanted impurities such as rock and dirt and comes in a mixture of different-sized fragments. Thus, another sequence of processes is necessary to make the coal consistent in quality and suitable for selling. These processes are called coal preparation or coal cleaning. Effective preparation of coal prior to combustion improves the homogeneity of the coal supplied, reduces transport costs, improves the burning efficiency, and produces fewer pollutants.

Coal preparation is the stage in coal production when the run-of-mine coal is processed into a range of clean, graded, and uniform coal products suitable for the commercial market, mostly power plants. In some cases, the run-of-mine coal is of such quality that it meets the user's specifications without the need for preparation, in which case the coal would merely be crushed and put through a large sievelike device to deliver the specified product.

A number of physical separation technologies are used in cleaning and preparing coal. After the raw run-of-mine coal is crushed, it is separated into various-sized fragments for optimum treatment. Larger material— lumps of coal about 10 to 150 mm (0.4 to 6 in) in length—is usually treated using a technology known as dense-medium separation. The dense medium is usually a liquid with a density just slightly greater than that of the coal. The coal can then be separated from other impurities, such as rock, by being floated in a tank containing the high-density liquid, which is usually a suspension of finely ground magnetite. Because the coal is lighter, it floats and is separated off, while heavier rock and other impurities sink and are removed as waste. Any magnetite mixed with the coal is separated using water sprays and is then recovered, using magnetic drums, and recycled.

The smaller-sized fragments are treated in a variety of ways. In the froth flotation method, coal particles are removed in a froth produced by blowing air into a water bath containing a chemical *reagent* (substance that takes part in a chemical reaction with another substance). The bubbles attract the coal but not the waste and are skimmed off to recover the smaller-sized fragments. After treatment, the smaller-sized fragments are screened and either dewatered or dried, and then recombined before going through final sampling and quality control procedures. Recombination also enables customers to selectively purchase different grades of coal. More expensive, higher quality supplies can be carefully mixed with lower quality coals to produce an average blend suited to the needs of commercial customers, such as power plants.

Once the coal has been extracted, it is moved from the mine to the power plant or other place of use or it is stored. Over short distances coal is usually transported by conveyor or truck. For long distances trains, barges, ships, or pipelines are used. Preventive measures are taken at every stage during transport and storage to reduce potential environmental impacts. Dust can be controlled by using water sprays, compacting the coal, and enclosing the stockpiles. Sealed systems can be used to move the coal from the stockpiles to the combustion plant. Well-designed coal-storage facilities can limit the problem of contaminated water run-off. All water is carefully treated before reuse or disposal.

ENVIRONMENTAL ISSUES

Because significant volumes of earth must be displaced to mine coal, coalmines and the resulting rock waste can harm the environment. burning coal releases environmentally harmful chemical compounds into the air.

Mining and Mining Waste

Surface mining has resulted in a great deal of damage to

the landscape. Many surface mines have removed acres of vegetation and altered topographic features, such as hills and valleys, leaving soil exposed for erosion. Longwall mining, which allows the mine to collapse, results in widespread land subsidence, or sinking.

Coal and rock waste, often dumped indiscriminately during surface and underground mining processes, weathers rapidly, producing *acid drainage.* Acid drainage contains sulfur-bearing compounds that combine with oxygen in water vapour to form sulfuric acid. In addition, weathering of coalmine waste can produce alkaline compounds, heavy metals, and sediments. Acid drainage, alkaline compounds, heavy metals, and sediment leached from mine waste into groundwater or washed away by rainwater can pollute streams, rivers, and lakes.

Today, enterprises in many countries must secure government permits before mining for coal. In the United States, mining companies must submit plans detailing proposed methods for blasting, road construction, land reclamation, and waste disposal. New land reclamation methods, driven by stringent laws and regulations, require coal mining companies to restore strip-mined landscapes to nearly premined conditions.

Burning Coal

The burning of coal produces environmentally harmful emissions. Some gases produced from burning coal, such as carbon dioxide, are known as greenhouse gases because they trap the Earth's heat like the roof of a greenhouse and may contribute to global warming. Other emissions from coal combustion can lead to air and water pollution.

PETROLEUM

Petroleum, or crude oil, naturally occurring oily, bituminous liquid composed of various organic chemicals. It is found in large quantities below the surface of Earth and is used as a fuel and as a raw material in the chemical industry. Modern industrial societies use it primarily to achieve a degree of mobility—on land, at sea, and in the air—that was barely imaginable less than 100 years ago. In addition, petroleum and its derivatives are used in the manufacture of medicines and fertilizers, foodstuffs, plastics, building materials, paints, and cloth and to generate electricity.

In fact, modern industrial civilization depends on petroleum and its products; the physical structure and way of life of the suburban communities that surround the great cities are the result of an ample and inexpensive supply of petroleum. In addition, the goals of developing countries—to exploit their natural resources and to supply foodstuffs for the burgeoning populations—are based on the assumption of petroleum availability.

In recent years, however, the worldwide availability of petroleum has steadily declined and its relative cost has increased. Many experts forecast that petroleum will no longer be a common commercial material by the mid-21st century. World Energy Supply.

Characteristics

The chemical composition of all petroleum is principally hydrocarbons, although a few sulfur-containing and oxygen-containing compounds are usually present; the sulfur content varies from about 0.1 to 5 percent. Petroleum contains gaseous, liquid, and solid elements. The consistency of petroleum varies from liquid as thin as gasoline to liquid so thick that it will barely pour. Small quantities of gaseous compounds are usually dissolved in the liquid; when larger quantities of these compounds are present, the petroleum deposit is associated with a deposit of natural gas.

Three broad classes of crude petroleum exist: the paraffin types, the asphaltic types, and the mixed-base types. The paraffin types are composed of molecules in which the number of hydrogen atoms is always two more than twice the number of carbon atoms. The characteristic molecules in the asphaltic types are naphthenes, composed of twice as many hydrogen atoms as carbon atoms. In the mixed-base group are both paraffin hydrocarbons and naphthenes.

Formation

Petroleum is formed under Earth's surface by the decomposition of marine organisms. The remains of tiny organisms that live in the sea—and, to a lesser extent, those of land organisms that are carried down to the sea in rivers and of plants that grow on the ocean bottoms—are enmeshed with the fine sands and silts that settle to the bottom in quiet sea basins. Such deposits, which are rich in organic materials, become the source rocks for the generation of crude oil.

The process began many millions of years ago with the development of abundant life, and it continues to this day. The sediments grow thicker and sink into the seafloor under their own weight. As additional deposits pile up, the pressure on the ones below increases several thousand times, and the temperature rises by several hundred degrees.

The mud and sand harden into shale and sandstone; carbonate precipitates and skeletal shells harden into limestone; and the remains of the dead organisms are transformed into crude oil and natural gas.

Once the petroleum forms, it flows upward in Earth's crust because it has a lower density than the brines that saturate the interstices of the shales, sands, and carbonate rocks that constitute the crust of Earth. The crude oil and natural gas rise into the microscopic pores of the coarser sediments lying. Frequently, the rising material encounters an impermeable shale or dense layer of rock that prevents migration; the oil has become trapped, and a reservoir of petroleum is formed. A significant amount of the upward-migrating oil, however, does not encounter impermeable rock but instead flows out at the surface of Earth or onto the ocean floor. Surface deposits also include bituminous lakes and escaping natural gas.

HISTORICAL DEVELOPMENT

These surface deposits of crude oil have been known to humans for thousands of years. In the areas where they occurred, they were long used for limited purposes, such as caulking boats, waterproofing cloth, and fueling torches. By the time the Renaissance began in the 14th century, some surface deposits were being distilled to obtain lubricants and medicinal products, but the real exploitation of crude oil did not begin until the 19th century.

The Industrial Revolution had by then brought about a search for new fuels, and the social changes it effected had produced a need for good, cheap oil for lamps; people wished to be able to work and read after dark. Whale oil, however, was available only to the rich, tallow candles had an unpleasant odor, and gas jets were available only in then-modern houses and apartments in metropolitan areas.

The search for a better lamp fuel led to a great demand for "rock oil"—that is, crude oil—and various scientists in the mid-19th century were developing processes to make commercial use of it. Thus British entrepreneur James Young, with others, began to manufacture various products from crude oil, but he later turned to coal distillation and the exploitation of oil shales. In 1852 Canadian physician and geologist Abraham Gessner obtained a patent for producing from crude oil a relatively clean-burning, affordable lamp fuel called kerosene; and in 1855 an American chemist, Benjamin Silliman, a report indicating the wide range of useful products that could be derived through the distillation of petroleum.

Thus the quest for greater supplies of crude oil began. For several years people had known that wells drilled for water and salt were occasionally infiltrated by petroleum, so the concept of drilling for crude oil itself soon followed.

The first such wells were dug in Germany from 1857 to 1859, but the event that gained world fame was the drilling of an oil well near Oil Creek, Pennsylvania, by "Colonel" Edwin L. Drake in 1859. Drake, contracted by the American industrialist George H. Bissell—who had also supplied Silliman with rock-oil samples for producing his report—drilled to find the supposed "mother pool" from which the oil seeps of western Pennsylvania were assumed to be emanating. The reservoir Drake tapped was shallow—only 21.2 m (69.5 ft) deep—and the petroleum was a paraffin type that flowed readily and was easy to distill.

Drake's success marked the beginning of the rapid growth of the modern petroleum industry. Soon petroleum received the attention of the scientific community, and coherent hypotheses were developed for its formation, migration upward through the earth, and entrapment. With the invention of the automobile and the energy needs brought on by World War I (1914-1918), the petroleum industry became one of the foundations of industrial society.

EXPLORATION

In order to find crude oil underground, geologists must search for a sedimentary basin in which shales rich in organic material have been buried for a sufficiently long time for petroleum to have formed. The petroleum must also have had an opportunity to migrate into porous traps that are capable of holding large amounts of fluid. The occurrence of crude oil in Earth's crust is limited both by these conditions, which must be met simultaneously, and by the time span of tens of millions to a hundred million years required for the oil's formation.

Petroleum geologists and geophysicists have many tools at their disposal to assist in identifying potential areas for drilling. Thus, surface mapping of outcrops of sedimentary beds makes possible the interpretation of subsurface features, which can then be supplemented with information obtained by drilling into the crust and retrieving cores or samples of the rock layers encountered.

In addition, increasingly sophisticated seismic techniques—the reflection and refraction of sound waves propagated through Earth—reveal details of the structure and interrelationship of various layers in the subsurface. Ultimately, however, the only way to prove that oil is present in the subsurface is to drill a well.

In fact, most of the oil provinces in the world have initially been identified by the presence of surface seeps, and most of the actual reservoirs have been discovered by so-called wildcatters who relied perhaps as much on intuition as on science. (The term *wildcatter* comes from West Texas, where in the early 1920s drilling crews encountered many wildcats as they cleared locations for exploratory wells. Shot wildcats were hung on the oil derricks, and the wells became known as wildcat wells.)

An oil field, once found, may comprise more than one reservoir—that is, more than one single, continuous, bounded accumulation of oil. Several reservoirs may be stacked one the other, isolated by intervening shales and impervious rock strata. Such reservoirs may vary in size from a few tens of hectares to tens of square kilometers, and from a few meters in thickness to several hundred or more. Most of the oil that has been discovered and exploited in the world has been found in a relatively few large reservoirs.

PRIMARY PRODUCTION

Most oil wells in the United States are drilled by the rotary method that was first described in a British patent in 1844 assigned to R. Beart. In rotary drilling, the drill string, a series of connected pipes, is supported by a derrick. The string is rotated by being coupled to the rotating table on the derrick floor. The drill bit at the end of the string is generally designed with three cone-shaped wheels tipped with hardened teeth. Drill cuttings are lifted continually to the surface by a circulating-fluid system driven by a pump.

Trapped crude oil is under pressure; were it not trapped by impermeable rock it would have continued to migrate upward, because of the pressure differential caused by its buoyancy, until it escaped at the surface of Earth. When a well bore is drilled into this pressured accumulation of oil, the oil expands into the low-pressure sink created by the well bore in communication with Earth's surface.

As the well fills up with fluid, however, a back pressure is exerted on the reservoir, and the flow of additional fluid into the well bore would soon stop, were no other conditions involved. Most crude oils, however, contain a significant amount of natural gas in solution, and this gas is kept in solution by the high pressure in the reservoir. The gas comes out of solution when the low pressure in the well bore is encountered, and the gas, once liberated, immediately begins to expand. This expansion, together with the dilution of the column of oil by the less dense gas, results in the propulsion of oil up to Earth's surface. Nevertheless, as fluid withdrawal continues from the reservoir, the pressure within the reservoir gradually decreases, and the amount of gas in solution decreases. As a result, the flow rate of fluid into the well bore decreases, and less gas is liberated. The fluid may not reach the surface, so a pump (artificial lift) must be installed in the well bore to continue producing the crude oil.

Eventually, the flow rate of the crude oil becomes so small, and the cost of lifting the oil to the surface becomes so great, that the well costs more to operate than the revenues that can be gained from selling the crude oil (after discounting the price for operating costs, taxes, insurance, and return on capital). The well's economic limit has then been reached and it is abandoned.

ENHANCED OIL RECOVERY

In primary production, no extraneous energy is added to the reservoir other than that required for lifting fluids from the producing wells. Most reservoirs are developed by numerous wells; and as primary production approaches its economic limit, perhaps only a few percent and no more than about 25 percent of the crude oil has been withdrawn from a given reservoir. The oil industry has developed methods for supplementing the production of crude oil that can be obtained mostly by taking advantage of the natural reservoir energy.

These supplementary methods, collectively known as enhanced oil recovery technology, can increase the recovery of crude oil, but only at the additional cost of supplying extraneous energy to the reservoir. In this way, the recovery of crude oil has been increased to an overall average of 33 percent of the original oil. Two successful supplementary methods are in use at this time: water injection and steam injection.

Water Injection

In a completely developed oil field, the wells may be

drilled anywhere from 60 to 600 m (200 to 2,000 ft) from one another, depending on the nature of the reservoir. If water is pumped into alternate wells in such a field, the pressure in the reservoir as a whole can be maintained or even increased.

In this way the rate of production of the crude oil also can be increased; in addition, the water physically displaces the oil, thus increasing the recovery efficiency. In some reservoirs with a high degree of uniformity and little clay content, water flooding may increase the recovery efficiency to as much as 60 percent or more of the original oil in place. Water flooding was first introduced in the Pennsylvania oil fields, more or less accidentally, in the late 19th century, and it has since spread throughout the world.

Steam Injection

Steam injection is used in reservoirs that contain very viscous oils, those that are thick and flow slowly. The steam not only provides a source of energy to displace the oil, but also causes a marked reduction in viscosity (by raising the temperature of the reservoir), so that the crude oil flows faster under any given pressure differential. This scheme has been used extensively in California and in the state of Zulia in Venezuela, where large reservoirs contain viscous oil. This technology is also being used to recover some of the vast accumulations of viscous crude oil, known as bitumen, along the Athabasca River in north central Alberta, Canada, and along the Orinoco River in eastern Venezuela.

Bitumen must be converted into a form of petroleum known as synthetic crude and is considered an unconventional crude oil. In Canada the steam technique most often used to extract bitumen is called steam-assisted gravity drainage (SAGD). This procedure requires a significant amount of energy and consequently is considered cost-efficient only when the price of conventional crude oil is high.

OFFSHORE DRILLING

Another method to increase oil-field production has been the construction and operation of offshore drilling rigs. The drilling rigs are installed, operated, and serviced on an offshore platform in water up to a depth of several hundred meters; the platform may either float or sit on legs planted on the ocean floor, where it is capable of resisting waves, wind, and—in Arctic regions—ice floes.

As in traditional rigs, the derrick is basically a device for suspending and rotating the drill pipe, to the end of which is attached the drill bit. Additional lengths of drill pipe are added to the drill string as the bit penetrates farther and farther into Earth's crust. The force required for cutting into the earth comes from the weight of the drill pipe itself.

To facilitate the removal of the cuttings, mud is constantly circulated down through the drill pipe, out through nozzles in the drill bit, and then up to the surface through the space between the drill pipe and the bore through the earth (the diameter of the bit is somewhat greater than that of the pipe). Successful bore holes have been drilled right on target, in this way, to depths of more than 6.4 km (more than 4 mi) from the surface of the ocean. Offshore drilling has resulted in the development of a significant additional reserve of petroleum—in the United States, about 5 percent of the total reserves.

REFINING

Once oil has been produced from an oil field, it is treated with chemicals and heat to remove water and solids, and the natural gas is separated. The oil is then stored in a tank, or battery of tanks, and later transported to a refinery by truck, railroad tank car, barge, or pipeline. Large oil fields all have direct outlets to major, common-carrier pipelines.

Basic Distillation

The basic refining tool is the distillation unit. In the United States after the Civil War (1861-1865), more than 100 still refineries were already in operation. Crude oil begins to vaporize at a temperature somewhat less than that required to boil water. Hydrocarbons with the lowest molecular weight vaporize at the lowest temperatures, whereas successively higher temperatures are required to distill larger molecules.

The first material to be distilled from crude oil is the gasoline fraction, followed in turn by naphtha and then by kerosene. The residue in the kettle, in the old still refineries, was then treated with caustic and sulfuric acid, and finally steam distilled thereafter. Lubricants and distillate fuel oils were obtained from the upper regions and waxes and asphalt from the lower regions of the distillation apparatus.

In the late 19th century the gasoline and naphtha fractions were actually considered a nuisance because little need for them existed. The demand for kerosene also began to decline because of the growing production of electricity and the use of electric lights. With the introduction of the automobile, however, the demand for gasoline suddenly burgeoned, and the need for greater supplies of crude oil increased accordingly.

Thermal Cracking

In an effort to increase the yield from distillation, the thermal cracking process was developed. In this process, the heavier portions of the crude oil were heated under pressure and at higher temperatures. This resulted in the large hydrocarbon molecules being split into smaller ones, so that the yield of gasoline from a barrel of crude oil was increased. The efficiency of the process was limited, however, because at the high temperatures and pressures that were used, a large amount of coke was deposited in the reactors.

This in turn required the use of still higher temperatures and pressures to crack the crude oil. A coking process was then invented in which fluids were recirculated; the process ran for a much longer time, with far less buildup of coke. Many refiners quickly adopted the process of thermal cracking.

Alkylation and Catalytic Cracking

Two additional basic processes, alkylation and catalytic cracking, were introduced in the 1930s and increased the gasoline yield from a barrel of crude oil. In alkylation small molecules produced by thermal cracking are recombined in the presence of a catalyst. This produces branched molecules in the gasoline boiling range that have superior properties—

example, higher antiknock ratings—as a fuel for high-powered engines such as those used in today's commercial airplanes.

In the catalytic-cracking process, the crude oil is cracked in the presence of a finely divided catalyst. This permits the refiner to produce many diverse hydrocarbons that can then be recombined by alkylation, isomerization, and catalytic reforming to produce high antiknock engine fuels and specialty chemicals.

The production of these chemicals has given birth to the gigantic petrochemical industry, which turns out alcohols, detergents, synthetic rubber, glycerin, fertilizers, sulfur, solvents, and the feedstocks for the manufacture of drugs, nylon, plastics, paints, polyesters, food additives and supplements, explosives, dyes, and insulating materials.

FUEL GASES

Fuel gases consist principally of hydrocarbons, that is, of molecular compounds of carbon and hydrogen. The properties of the various gases depend on the number and arrangement of the carbon and hydrogen atoms within their molecules. All these gases are odorless in the pure state, and carbon monoxide is toxic.

It is therefore common practice to add sulfur compounds to manufactured gas; such sulfur compounds, which are sometimes normally present in the gas, have an unpleasant smell and serve to give warning of a leak in the supply lines or gas appliance. In addition to their combustible components most gases have varying amounts of noncombustible nitrogen and water as their end products.

The devices used to burn gas for either heat or illumination consist of a burner nozzle and some means of mixing air with the gas before it reaches the nozzle in the Bunsen burner invented by British chemist and physicist Michael Faraday and improved and popularized by German chemist Robert Wilhelm Bunsen.

Fuel gases still in use are coal gas, made by the destructive distillation of coal producer gas and blast-furnace gas, made by the interaction of steam, air, and carbon; natural gas, drawn from gas deposits in the earth; and bottled gases, made from the lighter hydrocarbons.

Coal Gas

The most important coal-gasification processes aim chiefly at production of so-called pipeline quality gas, which is reasonably interchangeable with natural gas. Gas from coal, besides having pumping and heating specifications, must meet strict limits on content of carbon monoxide, sulfur, inert gases, and water. To meet these standards, most coal-gasification processes culminate with gas cleanup and methanation operations.

Various hydrogasification processes, in which hydrogen reacts directly with coal to form methane, are used today; these processes bypass the indirect step of producing synthesis gas, hydrogen and carbon monoxide, before an upgrading yields methane. Other coal-gas processes include the carbon dioxide acceptor process, employing the lime-bearing material dolomite, and the molten salt process. These processes work indirectly to produce synthesis gas first. Other gases manufactured formerly from coal and coke, such as illumination gas and coke-oven gas, are of little or no importance today.

PRODUCER AND BLAST-FURNACE GASES

Producer gas is a form of water gas, a term applied to steam-process gases. It is made by burning low-grade fuel (such as lignite or bituminous coal) in a closed vessel, called a producer, while passing a continuous stream of steam and air through the producer. Because of the air present in the producer, the resulting gas is approximately 50 percent incombustible nitrogen and is low in fuel value, having only about 28 percent the heating value of coke-oven gas.

Blast-furnace gas, which results from the interaction of limestone, iron ore, and carbon in blast furnaces, has some heating value because of its carbon monoxide content but contains about 60 percent nitrogen. Enormous quantities are produced during the operation of furnaces. Most of this gas is consumed in heating the air blast and driving the compressors for the blast. The heating value of blast-furnace gas is about 16 percent of that of coke-oven gas.

Natural Gas

A certain amount of natural gas almost always occurs in connection with oil deposits and is brought to the surface together with the oil when a well is drilled. Such gas is called casing-head gas. Certain wells, however, yield only natural gas.

Natural gas contains valuable organic elements that are important raw materials of the natural-gasoline and chemical industries. Before natural gas is used as fuel, hydrocarbons such as butane, propane, and natural gasoline are extracted as liquids. The remaining gas constitutes so-called dry gas, which is piped to domestic and industrial consumers for use as fuels; dry gas, devoid of butane and propane, also occurs in nature. Composed of the lighter hydrocarbons methane and ethane, dry gas is used also in the manufacture of plastics, drugs, and dyes.

Bottled Gas

Several of the lighter hydrocarbons, such as propane, butane, pentane, and mixtures of these gases are liquefied and employed as fuels. These so-called bottled gases, which are usually stored in steel cylinders, make possible the use of appliances such as cooking stoves and heaters in localities where a centralized gas supply is not available. Such bottled gases are produced from natural gas and petroleum.

Chapter 5

Energy from Waste

ENERGY FROM WASTE

Waste to energy conversion is an increasingly recognised approach to resolving two issues in one - waste management and sustainable energy. Waste represents an increasingly important fuel source. Using waste as fuel can have important environmental benefits. It can not only provide a safe and cost-effective way of waste disposal but can also help reduce carbon dioxide emissions.Whilst energy can be derived from waste by burning Landfill gas, there are also alternative methods to generate energy from waste.When waste is incinerated in large amounts, the heat energy can be recycled and used to heat factories, hospitals and other large buildings.

Alternatively, the heat can be used to generate electricity. This is done by using the steam created by combustion to drive a steam turbine.Electricity generating waste plants can typically process between 20,000 and 600,000 tonnes of waste per year, from which they can generate between 1 and 40 MW of electricity.Waste-derived fuel can also be burnt in boilers as an alternative to coal.Any energy that is recovered from biological waste can be regarded as renewable.

It comes from plant material. As plants grow they absorb carbon dioxide from the atmosphere. When biomass is used as fuel, this carbon dioxide is returned to the atmosphere, making the process carbon neutral.

Gasification

This is one of many technologies for converting waste into

energy. It is a thermo-chemical process in which the biomass is heated in an oxygen deficient atmosphere to produce a low energy gas which can then be used as a fuel in a turbine or combustion engine to produce electricity.

Strict environmental standards now apply in all European countries governing the emissions from energy from waste plants, particularly of heavy metals, furans and dioxins. All energy from waste plants must now meet these standards, which can be achieved through the installation of extensive state-of-the-art gas cleaning systems.

Biogas Generation

Biogas is a mixture comprising mainly methane and carbon dioxide. It is produced when organic matter decomposes in the abscence of oxygen.Biogas can be obtained directly from a variety of sources, including landfill, waste water and sewage treatment sites and other waste facilities. It can also be produced in dedicated anaerobic digestion facilities take in a wide variety of feedstocks, e.g. farm and food wastes to produce biogas.

Cattle, pigs and poultry all produce slurries that can be used to produce biogas. The slurries can be fermented in an anaerobic digester to produce a gas that is mainly methane and carbon dioxide. Some 40-60% of the organic matter present in the slurry is converted into biogas. After maturation, the remainder provides a stabilised residue that can be used as a soil conditioner.The gas can be used in gas engines to generate electricity or in boilers to provide process heat or space heating.Direct electricity generation from

- Landfill gas
- Sewage gas

Landfill Gas

Landfill gas is a mixture comprising mainly methane and carbon dioxide, formed when biodegradable wastes break down within a landfill as a result of anaerobic microbiological action.The biogas can be collected by drilling wells into the waste and extracting it as it is formed. It can then be used in an engine or turbine for power generation, or used to provide heat for industrial processes situated near the landfill site, such as in a brickworks.

Landfill sites can generate commercial quantities of landfill gas for up to 30 years after wastes have been deposited. Recovering this gas and using it as a fuel not only ensures the continued safety of the site after landfilling has finished, but also provides a significant long term income from power and/or heat sales.

Sewage Gas

Sewage gas is biogas produced by the digestion and incineration of sewage sludge. This biogas can be used to generate energy. As biogass is a source of energy with no net carbon emissions it has an important role to play in the reduction of greenhouse gases.

Sewage gas is a mixture of methane, carbon dioxide, nitrogen, hydrogen and hydrogen sulphide.The anaerobic digestion of sewage sludge involves fermenting it in tanks at a temperature between 32 and 36 degrees centigrade for about 25 days. The temperature in the digestion tank is kept constant using energy generated by a Combined Heat & Power unit.The biogas produced is then compressed and purified before a being transformed into mechanical and thermal energy. It can also power a generator to produce electricity.

Incineration

Incineration, the combustion of organic material such as waste, with energy recovery is the most common WtE implementation. Incineration may also be implemented without energy and materials recovery, but this is increasingly being banned in OECD (Organisation for Economic Co-operation and Development) countries.

Furthermore, all new WtE plants in OECD countries must meet strict emission standards. Hence, modern incineration plants are vastly different from the old types, some of which neither recovered energy nor materials. Modern incinerators reduce the volume of the original waste by 95-96 %, depending upon composition and degree of recovery of materials such as metals from the ash for recycling.

Concerns regarding the operation of incinerators include fine particulate, heavy metals, trace dioxin and acid gas emissions, even though these emissions are relatively low from modern incinerators. Other concerns include toxic fly ash and incinerator bottom ash (IBA) management. Waste resource ethics include the opinion that incinerators destroy valuable resources and the fear that they may reduce the incentives for recycling and waste minimization activities. Incinerators have electric efficiencies on the order of 14-28%. The rest of the energy can be utilized for e.g. district heating, but is otherwise lost as waste heat.

WtE technologies other than incineration

There are a number of other new and emerging technologies that are able to produce energy from waste and other fuels without direct combustion. Many of these technologies have the potential to produce more electric power from the same amount of fuel than would be possible by direct combustion. This is mainly due to the separation of corrosive components (ash) from the converted fuel, thereby allowing a higher combustion temperatures in e.g. boilers, gas turbines, internal combustion engines, fuel cells. Some are able to efficiently convert the energy into liquid or gaseous fuels:

Thermal technologies:

- Gasification (produces combustible gas, hydrogen, synthetic fuels)
- Thermal depolymerization (produces synthetic crude oil, which can be further refined)
- Pyrolysis (produces combustible tar/biooil and chars)
- Plasma arc gasification PGP or plasma gasification process (produces rich syngas including Hydrogen and Carbon Monoxide usable for fuel cells or generating electricity to drive the plasma arch, useable vitrified silicate and metal ingots, salt and sulphur)
- Gasplasma - A integrated process of front end gasification of shreaded waste to syngases, plasma arc

treatment of syngas and syngas cleaning and conditioning. Syngas, Plasmarok, Hydrogen, Carbon monoxide and CHP are the outputs. Residuals to landfill is less than 1%.

Non-thermal technologies:

- Anaerobic digestion (Biogas rich on methane)
- Ethanol production
- Mechanical biological treatment
  - MBT + Anaerobic digestion or Advanced MBT AMBT
  - MBT to Refuse derived fuel

Measurement of the Biomass Fraction of Waste

The biomass fraction of waste has a monetary value under multiple greenhouse gas protocols, such as the AB 32 programme in California and the Renewable Obligation Certificate programme in the United Kingdom. Biomass is considered to be carbon-neutral since the CO_2 liberated from the combustion of biomass is recycled in plants. The combusted biomass fraction of waste is used by waste to energy plants to reduce their overall reported CO_2 emissions.

Several methods have been developed by the European CEN 343 working group to determine the biomass fraction of waste fuels, such as Refuse Derived Fuel/Solid Recovered Fuel. The initial two methods developed were the manual sorting method and the selective dissolution method. Since each method suffered from limitations in properly characterizing the biomass fraction, two alternative methods have been developed.

One using the principles of radiocarbon dating. A technical review outlining the carbon 14 method was published in 2007. A technical standard of the carbon dating method will be published in 2008. In the United States, there is already an equivalent carbon 14 method under the standard method ASTM D6866.

The second method (so called balance method) employs existing data on materials composition and operating conditions of the incinerator and calculates the most probable result based on a mathematical-statistical model. Currently the balance method is installed at three Austrian incinerators.

Although carbon 14 dating can determine with some precision the biomass fraction of waste, it cannot determine directly the biomass calorific value. Determining the calorific value is important for green certificate programs such as the Renewable Obligation Certificate programme in the United Kingdom. These programs award certificates based on the energy produced from biomass.

Several research papers, including the one commissioned by the Renewable Energy Association in the UK, have been published that demonstrate how the carbon 14 result can be used to calculate the biomass calorific value. By contrast the balance method delivers all required information, namely, the ratio between biogenic and fossil energy production, as well as relative and total biogenic and fossil mass and carbon fractions. Moreover it requires no additional measurements and is therefore easy to install at low costs.

Energy Recycling

Energy recycling is utilizing energy that would normally be wasted, usually by converting it into electricity or thermal energy. Energy recycling — which can be undertaken at manufacturing facilities, power plants, and large institutions such as hospitals and universities — generally increases efficiency, thereby reducing energy costs and greenhouse gas pollution simultaneously. The process is noted for its potential to mitigate global warming profitably.

Forms of Energy Recycling

Waste heat recovery is a process that captures excess heat that would normally be discharged at manufacturing facilities and converts it into electricity and steam. A "waste heat recovery boiler" contains a series of water-filled tubes placed throughout the area where heat is released.

When high-temperature heat meets the boiler, steam is produced, which in turn powers a turbine that creates electricity. This process is similar to that of other fired boilers, but in this case, waste heat replaces a traditional flame. No fossil fuels are used in this process. Metals, glass, pulp and paper, silicon and other production plants are typical locations where waste heat recovery can be effective.

Combined heat and power (CHP), also called cogeneration, is, according to the U.S. Environmental Protection Agency, "an efficient, clean, and reliable approach to generating electricity and heat energy from a single fuel source. By installing a CHP system designed to meet the thermal and electrical base loads of a facility, CHP can greatly increase the facility's operational efficiency and decrease energy costs. At the same time, CHP reduces the emission of greenhouse gases, which contribute to global climate change." When electricity is produced on-site with a CHP plant, excess heat is recycled to produce both processed heat and additional power.

Current System

Both waste heat recovery and CHP constitute "decentralized" energy production, which is in contrast to traditional "centralized" power generated at large power plants run by regional utilities. The "centralized" system has an average efficiency of 34 percent, requiring about three units of fuel to produce one unit of power. By capturing both heat and power, CHP and waste heat recovery projects have higher efficiencies.

Department of Energy study found the potential for 135,000 megawatts of CHP in the U.S., and a Lawrence Berkley National Laboratory study identified about 64,000 megawatts that could be obtained from industrial waste energy, not counting CHP.

These studies suggest about 200,000 megawatts — or 20% — of total power capacity that could come from energy recycling in the U.S. Widespread use of energy recycling could therefore reduce global warming emissions by an estimated 20 percent. Indeed, as of 2005, about 42 percent of U.S. greenhouse gas pollution came from the production of electricity and 27 percent from the production of heat.

BIOMASS

Biomass, contraction for biological mass, the amount of living material provided by a given area of the earth's surface. Biomass energy, that is, the fuel energy that can be derived directly or indirectly from biological sources. Biomass energy from wood, crop residues, and dung remains the primary source of energy in developing regions. In a few instances it is also a major source of power, as in Brazil, where sugarcane is converted to ethanol fuel, and in China's Sichuan province, where fuel gas is obtained from dung.

Various research projects aim at development of biomass energy, but economic competition with petroleum has mainly kept such efforts at an early developmental stage.Biomass fuels work by releasing solar energy. Plants, through photosynthesis, convert solar energy to chemical energy, which fuels plant growth. People, in turn, use this stored solar energy through fuels such as wood, alcohol, and methane that are extracted from the plant life.

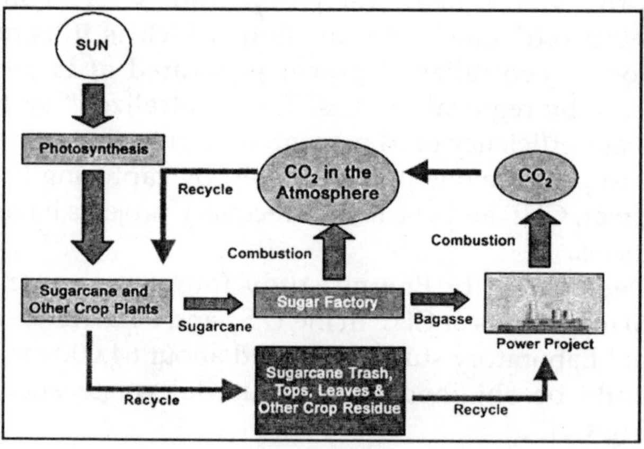

Fig. Biomass Energy.

BIOFUEL

Biofuel, fuel made from organic material produced by living things, in contrast to fossil fuels such as coal, petroleum, or natural gas that come from long-dead plants and microorganisms. Biofuel includes any solid, liquid, or gaseous fuel produced either directly from plants or indirectly from organic industrial, commercial, domestic, or agricultural wastes. In principle, burning biofuels adds less carbon to the environment than burning fossil fuels because the carbon atoms released by burning biofuel already existed as part of the modern carbon cycle. Burning fossil fuels, on the other hand, always adds extra carbon because the carbon they contain comes from a buried source that was not part of the modern carbon cycle.

Carbon dioxide is the main greenhouse gas thought to contribute to global warming. Biofuels are seen as one way to reduce the amount of carbon dioxide gas added to the atmosphere. Plants used for biofuels take up the same amount of carbon dioxide as they grow as was released by burning biofuels made from an earlier crop.

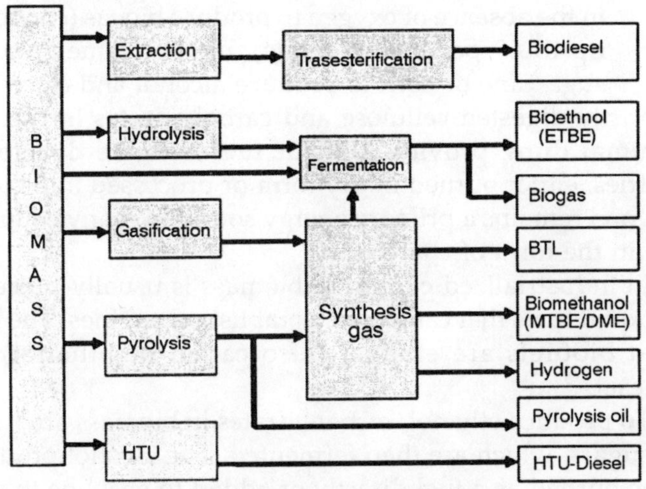

Fig. Biofuel Conversion Routes

Biofuels produced and used within the same country are a way to reduce dependence on foreign sources of oil and other fuels, providing energy security and an economic boost for agriculture and industry. Biofuels are also a type of renewable energy resource, unlike fossil fuels, which cannot be grown or created. Other renewable energy sources include solar energy and wind energy.

Types of Biofuels

Biofuels are mainly made from plant and plant-derived substances known as biomass. The energy in biomass originally came from the Sun through photosynthesis, a process that harnesses sunlight to turn carbon dioxide and water into sugars. Plants then use the sugars to create carbohydrates and cellulose that can later be turned into fuel.

There are three main methods for the development of biofuels:

- The burning of dry organic wastes (such as household refuse, industrial and agricultural wastes, straw, wood, and peat)
- Energy forestry (producing special fast-growing trees for wood that can be burned as fuel)
- The fermentation of wet wastes (such as animal dung) in the absence of oxygen to produce biogas (containing up to 60 percent methane), or the fermentation of sugarcane or corn to produce alcohol and esters.

Partly digested cellulose and carbohydrates in the form of animal dung provide a major fuel in some developing countries, either burned in dry form or processed into biogas. Firewood remains a primary energy source in many countries, often in the form of charcoal.

In industrialized countries, biomass is usually processed into liquid fuels that can power combustion engines. The main liquid biofuels are ethanol (also called bioethanol) and biodiesel.

To produce ethanol, carbohydrates in biomass are turned into sugars, which are then fermented. The alcohol produced can be burned as a fuel directly or added to gasoline to make gasohol. Starch crops such as corn and sugarcane that are rich in carbohydrates are common sources of ethanol. Other plant sources include wheat, rice, sorghum, sunflower, potatoes, and sugar beets. More complex processes are under study to convert cellulose into ethanol, allowing wood chips, fast-growing trees, or crops such as switch grass to be used.

Biodiesel is typically made from plant oils combined with alcohol to form esters. The esters can be burned as a fuel. In addition to vegetable oils from soybeans, rapeseed, or palm oil, biodiesel can also be made from used cooking oil, animal fats, or oils produced by certain types of microalgae.

Biomass can also be converted into bio-oil. Biomass plant material such as sawdust or sugarcane residue is heated to a temperature of around 400° to 500° C (752° to 932° F) in an oxygen-free environment for less than two seconds (a process called pyrolysis). The dark brown liquid that results can be burned as a fuel in boilers that generate electricity, though other uses may be feasible in the future.

Biofuel Use

The most widely used liquid biofuel in industrial countries is ethanol. In the United States in 2005, 3.9 billion gallons of ethanol were produced, mainly from corn. Most of that ethanol was blended with gasoline, accounting for 2.8 percent of total yearly gasoline sales. The European Union (EU) produced 718 million gallons of ethanol in 2005, mostly distilled from sugar beets and wheat grain.

The EU has set goals that 5.75 percent ethanol, derived from wheat, beets, potatoes, or corn, be added to fossil fuels by 2010 and 10 percent by 2020. Brazil is the world's largest producer of ethanol. About 15 percent of its liquid fuel is ethanol derived from sugarcane. Gasoline sold in Brazil is usually mixed with 25 percent ethanol. Many automobiles in the country have "flex-fuel" engines that can use either gasoline or pure ethanol.

Use of biodiesel is also increasing, often blended with diesel derived from petroleum. In 2005 the EU produced 858 million gallons of biodiesel, primarily from rapeseed oil. The United States produced about 250 million gallons of biodiesel in 2006, nearly all of it from soybean oil.

Potential Negative Effects of Biofuel Production

Politicians, scientists, and economists have promoted biofuels as renewable energy sources that add less carbon to the environment and provide energy independence. However, some experts have raised concerns about possible negative consequences of growing crops to provide biofuels. Food prices may rise if a major percentage of grain crops are grown for energy and if areas once used to grow food are converted to energy crops.

The price increases could affect impoverished populations in developing countries, in particular. Rain forests and other tropical areas could be cleared to cultivate crops such as oil palm and soybeans, at the same time displacing populations in regions such as Southeast Asia and the Amazon.

BIOENERGY

Socio-economic development, standard of living, as well as the quality of life of any country largely and primarily depend on the availability and supply of energy. The availability and supply of energy come under great pressure because of increasing demands of the ever-increasing global population.

This eventually leads to the problem of *energy crisis*. This is especially true for the developing countries like India. At present, the main conventional source of energy used all over the world is the 'fossil fuels' (i.e. the huge deposits of plant and animal remains accumulated during the geological past and which serve as the main source of energy such as. petroleum (oil), natural gas, coal, derivatives of coal, oil shale, tar sands, etc.).

However, fossil fuels are non-renewable and hence their deposits will eventually be exhausted. Therefore, alternative energy sources must be tapped and exploited to their full potential. Solar energy, nuclear energy, energy from tides, energy from wind, and hydroelectric energy are some of the alternative resources. However, one of the major renewable alternative sources is the bioenergy obtained from biomass.

Biofuels are a renewable resource of energy. The primary source of the energy stored in biomass is solar energy which is trapped by green plants during photosynthesis and stored as potential energy in organic molecules. Biofuels are obtained from the terrestrial biomass (wood, forest litter etc.), aquatic biomass (algae and water weeds like water hyacinth, *Pistia, Salvinia, Azolla,* etc.), various types of organic wastes (agricultural and industrial wastes, municipal and domestic wastes like sewage, garbage, animal drug, wastes from slaughter houses, etc.). Plants that produce alcohol, oil and petroleum (petro-plants) also are a source of biofuels.

Various methods and technologies which convert biomass into bioenergy fuels fall into two main categories.

- The non-biological processes such as
  - Direct combustion
  - Conversion of biomass into liquid fuels such as fuel oil (e.g. by pyrolysis - a type of fertilization, liquefaction, etc.) and
  - Gassification..
- The biological process or bioconversion involves the conversion of biomass into bioenergy, fertilizer, food and chemicals through the biological action of micro-organisms. One of the important biofuels obtained through bio-conversion is natural gas, a biogas (methane).

COMMON METHODS OF BIOGAS FORMATION

Methane is the main constituent (63%) of the biogas. (The other major constituent is carbon dioxide = 30%.) It is known by various names such as biofuel, sewerage gas, Klar gas, sludge gas, will-o-the wisp of marsh lands, fool's fire, gobar gas, bio-energy and even fuel of the future. Biogas is often used for cooking and lighting purposes in the rural sector. It burns with a blue flame without smoke and is without odor.

Biogas is formed under anaerobic conditions when organic materials are converted into gases (fuel) and organic fertilizer (sludge) through microbial reactions. Similarly, fuel oil is produced from domestic wastes, agricultural wastes, forestry wastes, etc. by the process of pyrolysis (destructive distillation or decomposition of organic wastes in anaerobic conditions at high temperatures).

RAW MATERIALS AND SUBSTRATES

Biogas can be produced from the biomass of plant origin (either terrestrial biomass or aquatic biomass) or animal origin (cattle dung, domestic sewage, organic manure, poultry, slaughter house and fishery wastes, etc.) An enormous quantity of sugarcane wastes is produced in the sugar mills. Sugarcane bagasse and molasses are the two important wastes.

Bagasse is the cellulosic material of sugarcane produced after the extraction of the sugar juice. It is used for the production of biogas, alcohols, single cell protein (SCP), etc. Sugarcane molasses contains about 50-55% fermentable sugars and hence is an important source for the production of liquid fuel, alcoholic beverages and animal feed.Similarly, straw, agricultural wastes, forest litter, aquatic weeds like water hyacinth, Pistia, Salvinia, etc., and cyanobacteria are also the common raw materials used for the production of biogas.

PRODUCER GAS

Producer gas is a mixture of approximately 25% carbon monoxide, 55% nitrogen, 13% hydrogen and 7% other gases.

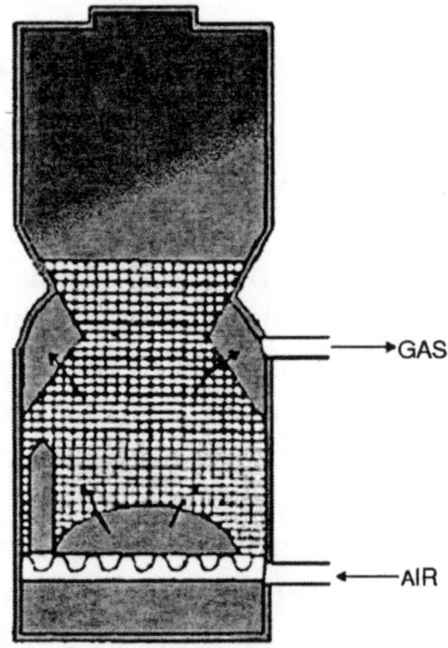

Fig. Gas Generator

It is obtained by burning coal or coke in the generators with a restricted supply of air, or by passing air and steam through a bed of red hot fuel. Producer gas is cheap and used as a fuel mainly in glass furnaces and metallurgical furnaces. It also serves as a fuel in gas engines to operate tractors, motor cars and truckso. It is also used as a source of nitrogen for the preparation of ammonia.However, producer gas is highly flammable. It is also toxic and explosive.

METHANE

The common raw materials used for the production of methane are biomass products such as human and animal wastes, agricultural, industrial and domestic residues. Cattle dung is the most common material used in most biogas plants.

Stage Process involved	*Group of bacteria involved*
Solubilization	Hydrolytic fermentor bacteria (mostly anaerobic).
Acetogenesis	Hydrogen - producing acetogenic bacteria (facultative anaerobic)
Methanogenesis	Methanogens(i.e. methane producing anaerobic bacteria)

Micro-organisms involved

Conversion of the biomass into biogas (methane) is a process called anaerobic digestion. It is completed in three stages. Breakdown of the biomass as each stage is done by a different group of bacteria.

GOBAR GAS PLANT

As the biogas (gobar gas) production is an anaerobic process, it is carried out in an air tight, closed cylindrical concrete tank called a digester. In countries like India, domestic gobar gas plants use single digester tanks.

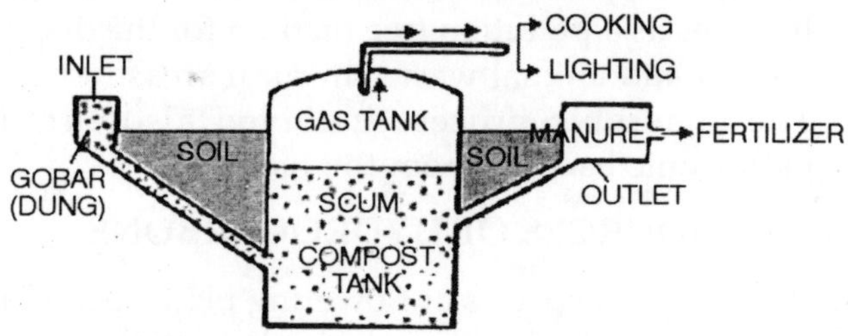

Fig. Domestic Gobar gas plant (Diagrammatic)

The tank has a concrete inlet basin on one side for feeding

fresh cattle dung (*gobar*). There is a concrete outlet on the outer side for removing the digested sludge. The top of the tank serves as the gas tank. It has an outlet pipe for the gobar gas/ biogas.

Process

Fresh cattle dung is fed into the digester tank through the inlet and allowed to remain there. After about 50 days, a sufficient amount of gobar gas is accumulated in the gas tank. It is conducted through the outlet pipe and used for domestic purposes. The digested sludge (digested biomass) is removed from the tank and is used as fertilizer. The tank is again filled with fresh dung and the process is repeated.

Uses

- Gobar gas is mainly used for cooking and lighting in rural areas.
- Also used in internal combustion engines to power water pumps and electric generators.
- Used as a fuel in the fuel type refrigerators.

There are certain advantages in using methane (gobar gas) as a source of energy:

- Methane is very insoluble, hence it separates very readily from the fermentor system.
- It can be very easily collected, pressurized or liquified for storage.
- It is readily combustible.

Other advantages are:

- Sludge is used as fertilizer.
- It serves as a pollution free method for the disposal of human and animal wastes in rural areas.
- The anaerobic process involved also eliminates pathogenic bacteria from the biomass.

PLANTS AS SOURCES OF HYDROCARBONS

There are certain species of flowering plants belonging to different families which convert a substantial amount of photosynthetic products into latex. The latex of such plants contains liquid hydrocarbons of high molecular weight (10,000da). These hydrocarbons can be converted into high grade transportation fuel (i.e. petroleum).

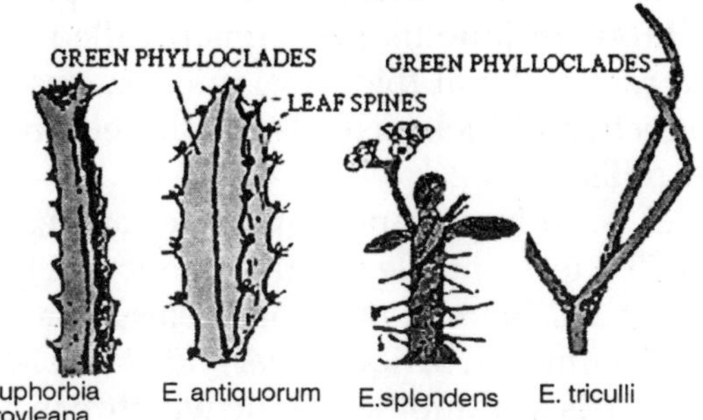

Fig. Some Latex-containing Species of Euphorbis (Petro-plants)

Therefore, hydrocarbon producing plants are called petroleum plants or petroplants and their crop as petrocrop. Natural gas is also one of the products obtained from hydrocarbons. Thus, petroleum plants can be an alternative source for obtaining petroleum to be used in diesel engines.

Normally, some of the latex-producing plants of families Euphorbiaceae, Apocynaceae, Asclepiadaceae, Sapotaceae, Moraceae, Dipterocarpaceae, etc. are petroplants. Similarly, sunflower (family Composiae), Hardwickia pinnata (family Leguminosae) are also petroplants. Some algae also produce hydrocarbons.

Euphorbia

Different species of Euphorbia of family Euphorbiaceae serve as the petroplants. The hydrocarbons from plants of Euphorbiaceae. He suggested that these can be the renewable substitute for the conventional petroleum sources. Latex of Euphorbia lathyrus contains fairly high percentage of terpenoids. These can be converted into high grade transportation fuel. Similarly the carbohydrates (hexoses) from such plants can be used for ethanol formation.

Sugarcane and Sugar Beet

Sugarcane (Saccharum officinarum - family Gramineae)

is the main source of raw material for sugar industry. The wastes from sugar industry include bagasse, molasses and press mud.

After extracting the cane juice for sugar production, the cellulosic fibrous residue that remains is called bagasse. It is used as the raw material (biomass) and processed variously for the production of fuel, alcohols, single cell protein as well as in paper mills.

Molasses is an important by-product of sugar mills and contains 50-55% fermentable sugars. One ton of molasses can produce about 280 liters of ethanol. Molasses is used for the production of animal feed, liquid fuel and alcoholic beverages.

Sugar beet is yet another plant which contains a high percentage of sugars stored in fleshy storage roots. It is also an important source for production of sugar as well as ethanol. Raising crops like sugarcane, sugar beet, tapioca, potato, maize etc. purely for production of ethanol is described as energy cropping, and the crops are called energy crops. Cultivation of plants (trees) for obtaining fuel/fire wood is described as energy plantation.

GASOHOL

Gasohol, a blend of nine parts unleaded gasoline and one part alcohol (ethanol or methanol), used extensively in some countries to reduce the cost of gasoline as automobile fuel. Raw materials for methanol production are coal and organic wastes, especially waste-wood products, while ethanol may be distilled from grain, sugar crops, or almost any starchy plant. Although in most of the world in the early 1980s it remained more expensive than gasoline, gasohol is derived from renewable sources.

Alcohol was used interchangeably with gasoline in the first internal-combustion engines in the 1870s, and gasoline-alcohol blends have been used periodically in Europe when petroleum was in short supply. Two gasoline-alcohol blends, alcoline and agrol, were sold in the U.S. in the 1930s but were unable to compete successfully with low-cost gasoline. The oil shortages of the 1970s prompted a revival of interest in alcohol blends, an interest increased in 1985 by the proposed banning of leaded gasolines by the end of the 1980s. Gasohol can be used without modifying the carburetor, ignition timing, or fuel lines of an automobile and provides a slightly higher octane than unleaded regular gasoline.

GASOLINE

Gasoline, mixture of the lighter liquid hydrocarbons used chiefly as a fuel for internal-combustion engines. It is produced by the fractional distillation of petroleum; by condensation or adsorption from natural gas; by thermal or catalytic decomposition of petroleum or its fractions; by the hydrogenation of producer gas or coal; or by the polymerization of hydrocarbons of lower molecular weight.

Gasoline produced by the direct distillation of crude petroleum is known as straight-run gasoline. It is usually distilled continuously in a bubble tower, which separates the gasoline from the other fractions of the oil having higher boiling points, such as kerosene, fuel oil, lubricating oil, and grease. The range of temperatures in which gasoline boils and is distilled off is roughly between 38° and 205° C (100° and 400° F). The yield of gasoline from this process varies from about 1 percent to about 50 percent, depending on the petroleum. Straight-run gasoline now makes up only a small part of U.S. gasoline production because of the superior merits of the various cracking processes.

In many parts of the country natural gas contains a percentage of natural gasoline that may be recovered by condensation or adsorption. The most common process for the extraction of natural gasoline includes passing the gas as it comes from the well through a series of towers containing a light oil called straw oil. The oil absorbs the gasoline, which is then distilled off. Other processes involve adsorption of the gasoline on activated alumina, activated carbon, or silica gel.

High-grade gasoline can be produced by a process known as hydrofining, that is, the hydrogenation of refined petroleum oils under high pressure in the presence of a catalyst such as molybdenum oxide. Hydrofining not only converts oils of low value into gasoline of higher value but also at the same time purifies the gasoline chemically by removing undesirable elements such as sulfur. Producer gas, coal, and coal-tar distillates can also be hydrogenated to form gasoline.

For use in high-compression engines, it is desirable to produce gasoline that will burn evenly and completely in order to prevent knocking, the sound and damage caused by premature ignition of a part of the fuel and air charge in the combustion chamber of an internal-combustion engine. The antiknock characteristics of a gasoline are directly related to its efficiency and are indicated by its octane number. This is a rating that describes performance of a fuel in comparison with that of a standard fuel containing given percentages of isooctane and heptane.

If the performance of the rated fuel is the same as that of a standard fuel with a certain percentage of isooctane, the octane number given the rated fuel is the same as the percentage of isooctane in the standard fuel. The higher this number, the less likely a fuel is to cause knocking. Cracked gasoline has better antiknock characteristics than straight-run gasoline, and any gasoline can be improved by the addition of such substances as tetraethyl or tetramethyl lead. Since it was discovered, however, that the emission of lead from gasolines combined with such additives is dangerous to living beings—among other effects, raising blood pressure—research on new ways to reduce the knocking characteristics of gasoline was intensified.

WORLD ENERGY SUPPLY

World Energy Supply, combined resources by which the nations of the world attempt to meet their energy needs. Energy is the basis of industrial civilization; without energy, modern life would cease to exist. During the 1970s the world began a painful adjustment to the vulnerability of energy supplies. In the long run, conserving energy resources may provide the time needed to develop new sources of energy, such as hydrogen fuel cells, or to develop alternative energy sources, such as solar energy and wind energy. While this development occurs, however, the world will continue to be vulnerable to disruptions in the supply of oil, which, after World War II (1939-1945), became the most favored energy source.

BACKGROUND OF TODAY'S SITUATION

Wood was the first and, for most of human history, the major source of energy. It was readily available, because extensive forests grew in many parts of the world and the amount of wood needed for heating and cooking was relatively modest. Certain other energy sources, found only in localized areas, were also used in ancient times: asphalt, coal, and peat from surface deposits and oil from seepages of underground deposits.

This situation changed when wood began to be used during the Middle Ages to make charcoal. The charcoal was heated with metal ore to break up chemical compounds and free the metal. As forests were cut and wood supplies dwindled at the onset of the Industrial Revolution in the mid-18th century, charcoal was replaced by coke (produced from coal) in the reduction of ores. Coal, which also began to be used to drive steam engines, became the dominant energy source as the Industrial Revolution proceeded.

Growth of Petroleum Use

Although for centuries petroleum (also known as crude oil) had been used in small quantities for purposes as diverse as medicine and ship caulking, the modern petroleum era began when a commercial well was brought into production in Pennsylvania in 1859. The oil industry in the United States expanded rapidly as refineries sprang up to make oil products from crude oil. The oil companies soon began exporting their principal product, kerosene—used for lighting—to all areas of the world. The development of the internal-combustion engine and the automobile at the end of the 19th century created a vast new market for another major product, gasoline. A third major product, heavy oil, began to replace coal in some energy markets after World War II.

The major oil companies, which are based principally in the United States, initially found large oil supplies in the United States. As a result, oil companies from other countries—especially Britain, the Netherlands, and France—began to search for oil in many parts of the world, especially the Middle East. The British brought the first field there into production just before World War I (1914-1918).

During World War I, the U.S. oil industry produced two-thirds of the world's oil supply from domestic sources and imported another one-sixth from Mexico. At the end of the war and before the discovery of the productive East Texas fields in 1930, however, the United States, with its reserves strained by the war, became a net oil importer for a few years.

During the next three decades, with occasional federal support, the U.S. oil companies were enormously successful in expanding in the rest of the world. By 1955 the five major U.S. oil companies produced two-thirds of the oil for the world oil market (not including North America and the Soviet bloc).

Two British-based companies produced almost one-third of the world's oil supply, and the French produced a mere one-fiftieth. The next 15 years were a period of serenity for energy supplies. The seven major U.S. and British oil companies provided the world with increasing quantities of cheap oil. The world price was about a dollar a barrel, and during this time the United States was largely self-sufficient, with its imports limited by a quota.

Formation of OPEC

Two series of events coincided to change this secure supply of cheap oil into an insecure supply of expensive oil. In 1960, enraged by unilateral cuts in oil prices by the seven big oil companies, the governments of the major oil-exporting countries formed the Organization of Petroleum Exporting Countries, or OPEC.

OPEC's goal was to try to prevent cuts in the price that the member countries—Venezuela and four countries around the Persian Gulf—received for oil. They succeeded, but for a decade they were unable to raise prices. In the meantime, increasing oil consumption throughout the world, especially in Europe and Japan, where oil displaced coal as a primary source of energy, caused an enormous expansion in the demand for oil products.

THE ENERGY CRISIS

The year 1973 brought an end to the era of secure, cheap oil. In October, as a result of the Arab-Israeli War, the Arab oil-producing countries cut back oil production and embargoed oil shipments to the United States and the Netherlands. Although the Arab cutbacks represented a loss of less than 7 percent in world supply, they created panic on the part of oil companies, consumers, oil traders, and some governments.

Wild bidding for crude oil ensued when a few producing nations began to auction off some of their oil. This bidding encouraged the OPEC nations, which now numbered 13, to raise the price of all their crude oil to a level as high as eight times that of a few years earlier. The world oil scene gradually calmed, as a worldwide recession brought on in part by the higher oil prices trimmed the demand for oil. In the meantime, most OPEC governments took over ownership of the oil fields in their countries.

In 1978 a second oil crisis began when, as a result of the revolution that eventually drove the Shah of Iran from his throne, Iranian oil production and exports dropped precipitously. Because Iran had been a major exporter, consumers again panicked. A replay of 1973 events, complete with wild bidding, again forced up oil prices during 1979. The outbreak of war between Iran and Iraq in 1980 gave a boost to oil prices. By the end of 1980 the price of crude oil stood at 19 times what it had been just ten years earlier.

The very high oil prices again contributed to a worldwide recession and gave energy conservation a big push. As oil demand slackened and supplies increased, the world oil market slumped. Significant increases in non-OPEC oil supplies, such as those in the North Sea, Mexico, Brazil, Egypt, China, and India, pushed oil prices even lower. Production in the Soviet Union reached 11.42 million barrels per day by 1989, accounting for 19.2 percent of world production in that year.

Despite the low world oil prices that have prevailed since 1986, concern over disruption has continued to be a major focus of energy policy in the industrialized countries. The short-term increases in prices following Iraq's invasion of Kuwait in 1990 reinforced this concern. Owing to its vast reserves, the Middle East will continue to be the major source of oil for the foreseeable future. However, new discoveries in the Caspian Sea region suggest that countries such as Kazakhstan may become major sources of petroleum in the 21st century.

Current Status

In the 1990s, oil production by non-OPEC countries remained strong and production by OPEC countries rebounded. The result at the end of the 20th century was a world oil surplus and prices (when adjusted for inflation) that were lower than in 1972. Experts are uncertain about future oil supplies and prices. Low prices have spurred greater oil consumption, and experts question how long world petroleum reserves can keep pace with increased demand.

Many of the world's leading petroleum geologists believe the world oil supply will peak around 80 million barrels per day between 2010 and 2020. On the other hand, many economists believe that even modestly higher oil prices might lead to greater supply, since the oil companies would then have the economic incentive to exploit less accessible oil deposits.

Natural gas may be increasingly used in place of oil for applications such as power generation and transportation. One reason is that world reserves of natural gas have doubled since 1976, in part because of the discovery of major deposits of natural gas in Russia and in the Middle East. New facilities and pipelines are being constructed to help process and transport this natural gas from production wells to consumers.

PETROLEUM AND NATURAL GAS

Petroleum (crude oil) and natural gas are found in

commercial quantities in sedimentary basins in more than 50 countries in all parts of the world. The largest deposits are in the Middle East, which contains more than half the known oil reserves and almost one-third of the known natural-gas reserves. The United States contains only about 2 percent of the known oil reserves and 3 percent of the known natural-gas reserves.

Drilling

Geologists and other scientists have developed techniques that indicate the possibility of oil or gas being found deep in the ground. These techniques include taking aerial photographs of special surface features, sending shock waves through the earth and reflecting them back into instruments, and measuring the earth's gravity and magnetic field with sensitive meters. Nevertheless, the only method by which oil or gas can be found is by drilling a hole into the reservoir. In some cases oil companies spend many millions of dollars drilling in promising areas, only to find dry holes.

For a long time, most wells were drilled on land, but after World War II drilling commenced in shallow water from platforms supported by legs that rested on the sea bottom. Later, floating platforms were developed that could drill at water depths of 1,000 m (3,300 ft) or more. Large oil and gas fields have been found offshore: in the United States, mainly off the Gulf Coast; in Europe, primarily in the North Sea; in Russia, in the Barents Sea and the Kara Sea; and off Newfoundland and Labrador, and Brazil. Most major finds in the future may be offshore.

Production

As crude oil or natural gas is produced from an oil or gas field, the pressure in the reservoir that forces the material to the surface gradually declines. Eventually, the pressure will decline so much that the remaining oil or gas will not migrate through the porous rock to the well. When this point is reached, most of the gas in a gas field will have been produced, but less than one-third of the oil will have been extracted. Part of the remaining oil can be recovered by using water or carbon dioxide gas to push the oil to the well, but even then, one-fourth to one-half of the oil is usually left in the reservoir.

In an effort to extract this remaining oil, oil companies have begun to use chemicals to push the oil to the well, or to use fire or steam in the reservoir to make the oil flow more easily. New techniques that allow operators to drill horizontally, as well as vertically, into very deep structures have dramatically reduced the cost of finding natural gas and oil supplies.

Crude oil is transported to refineries by pipelines, barges, or giant oceangoing tankers. Refineries contain a series of processing units that separate the different constituents of the crude oil by heating them to different temperatures, chemically modifying them, and then blending them to make final products. These final products are principally gasoline, kerosene, diesel oil, jet fuel, home heating oil, heavy fuel oil, lubricants, and feedstocks, or starting materials, for petrochemicals.

Natural gas is transported, usually by pipelines, to customers who burn it for fuel or, in some cases, make petrochemicals from chemicals extracted, or "stripped," from it. Natural gas can be liquefied at very low temperatures and transported in special ships. This method is much more costly than transporting oil by tanker. Oil and natural gas compete in a number of markets, especially in generating heat for homes, offices, factories, and industrial processes.

Pollution Problems

In its early days, the oil industry generated considerable environmental pollution. Through the years, however, under the dual influences of improved technology and more stringent regulations, it has become much cleaner.

The effluents from refineries have decreased greatly and, although well blowouts still occur, new technology has tended to make them relatively rare. The policing of the oceans, on the other hand, is much more difficult. Oceangoing ships are still a major source of oil spills. In 1990 the Congress of the

United States passed legislation requiring tankers to be double hulled by the end of the decade.

Another source of pollution connected with the oil industry is the sulfur in crude oil. Regulations of national and local governments restrict the amount of sulfur dioxide that can be discharged by factories and utilities burning fuel oil. Because removing sulfur is expensive, however, regulations still allow some sulfur dioxide to be discharged into the air.

Many scientists believe that another potential environmental problem from refining and burning large amounts of oil and other fossil fuels (such as coal and natural gas) occurs when carbon dioxide (a by-product of the burning of fossil fuels), methane (which exists in natural gas and is also a by-product of refining petroleum), and other by-product gases accumulate in the atmosphere. These gases are known as greenhouse gases, because they trap some of the energy from the Sun that penetrates Earth's atmosphere. This energy, trapped in the form of heat, maintains Earth at a temperature that is hospitable to life. Certain amounts of greenhouse gases occur naturally in the atmosphere.

However, the immense quantities of petroleum, coal, and other fossil fuels burned during the world's rapid industrialization over the last 200 years are a contributing source of higher levels of carbon dioxide in the atmosphere. During that time period, these levels have increased by about 28 percent. This increase in atmospheric carbon dioxide, coupled with the continuing loss of the world's forests (which absorb carbon dioxide), has led many scientists to predict a rise in global temperature. This increase in global temperature might disrupt weather patterns, disrupt ocean currents, lead to more violent storms, and create other environmental problems.

In 1992 representatives of over 150 countries convened in Rio de Janeiro, Brazil, and agreed on the need to reduce the world's emissions of greenhouse gases. In 1997 world delegations again convened, this time in Kyôto, Japan. During the Kyôto meeting, representatives of 160 nations, including the United States, signed an agreement known as the Kyôto

Protocol, which would require 38 industrialized nations to limit emissions of greenhouse gases to levels that are an average of 5 percent below the emission levels of 1990.

In order to reduce their fossil fuel emissions to achieve these levels, the industrialized nations would have to shift their energy mix toward energy sources that do not produce as much carbon dioxide, such as natural gas, or to alternative energy sources, such as hydroelectric energy, solar energy, wind energy, or nuclear energy.

While the governments of some industrialized nations have ratified the Kyôto Protocol, others have not. A major blow to the protocol came in March 2001 when United States president George W. Bush rejected it, saying it would damage the U.S. economy. Under the previous administration of President Bill Clinton, the United States had volunteered to reduce greenhouse gas emissions to 7 percent below 1990 levels. Bush's rejection meant that the world's largest consumer of fossil fuels would not participate in the Kyôto Protocol.

Reserves

Oil shale, heavy oil deposits, and tar sands are the most prevalent forms of petroleum found in the world. Reserves of these sources are many times more abundant than the world's total known reserves of crude oil. Because of the high cost of converting shale oil and tar sands into usable petroleum products, however, only a small percentage of the available material is processed commercially.

An industry to make oil products from tar sands has been started in Canada, and Venezuela is looking at the prospects of developing the vast reserves of tar sands in its Orinoco River basin. Nevertheless, the quantity of oil products produced from these two raw materials is small compared with the total production of conventional crude oil.

Until world petroleum prices increase, the quantity of oil produced from oil shale and tar sands will likely remain small relative to the production of conventional crude oil.Coal is a general term for a wide variety of solid materials that are high in carbon content. Most coal is burned by electric utility companies to produce steam to turn their generators. Some coal is used in factories to provide heat for buildings and industrial processes. A special, high-quality coal is turned into metallurgical coke for use in making steel.

Current Trends

In industrialized countries, the greater convenience and lower costs of oil and gas in the earlier 20th century virtually forced coal out of the market for heating homes and offices and driving locomotives. Oil and gas also ate heavily into the industrial market for coal. Only an expanding utility market enabled coal output to remain relatively constant between 1948 and 1973. Even in the utility market, as oil and gas captured a greater share, coal's contribution to the total energy picture dropped dramatically—in the United States, for instance, from about one-half to less than one-fifth.

The dramatic jumps in oil prices after 1973, however, gave coal a major cost advantage for utilities and large industrial customers, and coal began to recapture some of its lost markets. In contrast to the industrialized countries, developing countries that have large coal reserves continue to use coal for industrial and heating purposes.

The average price of coal has remained virtually unchanged since the early 1980s and is forecast to decline in the early part of the 21st century. However, in industrialized countries the need to comply with stricter environmental regulations has made burning coal more costly.

POLLUTION PROBLEMS

Despite coal's relative cheapness and huge reserves, the growth in the use of coal since 1973 has been much less than expected, because coal is associated with many more environmental problems than is oil. Underground mining can result in black lung disease for miners, the sinking of the land over mines, and the drainage of acid into water tables.

Surface mining requires careful reclamation, or the unrestored land will remain scarred and unproductive. In addition, the burning of coal causes emission of sulfur dioxide particles, nitrogen oxide, and other impurities. Acid rain—rainfall and other forms of precipitation with a relatively high acidity that is damaging lakes and some forests in many regions—is believed to be caused in part by such emissions. The U.S. Clean Air Act of 1970 (revised in 1970 and 1990) provides the federal legal basis for controlling air pollution. This legislation has significantly reduced emissions of sulfur oxides—known as acid gases.

The Clean Air Act requires facilities such as coal-burning power plants to burn low-sulfur coal. In the 1990s concern over the possible warming of the planet as a result of the greenhouse effect caused many governments to consider policies to reduce the carbon dioxide emissions produced by burning coal, oil, and natural gas. During the world's rapid industrialization through the 19th and 20th centuries, levels of carbon dioxide in the atmosphere increased approximately 28 percent from preindustrial levels.

Solving these problems is costly, and who should pay is a matter of controversy. As a result, coal consumption may continue to grow more slowly than would otherwise be expected. The vast coal reserves, the improved technologies to reduce pollution, and the development of coal gasification still indicate, however, that the market for coal will increase in coming years.

SYNTHETIC FUELS

Synthetic fuels do not occur in nature but are made from natural materials. Gasohol, is a mixture of gasoline and alcohol made from sugars produced by living plants. Although making various types of fuel from coal is possible, the large-scale production of fuel from coal will likely be limited by high costs and pollution problems, some of which are not yet known. The manufacture of alcohol fuels in large quantities will likely be restricted to regions, such as parts of Brazil, where a combination of low-cost labor and land, plus a long growing season, make it economical. Thus, synthetic fuels are unlikely to make an important contribution to the world's energy supply anytime soon.

NUCLEAR ENERGY

Nuclear energy is generated by the splitting, or fissioning, of atoms of uranium or heavier elements. The fission process releases heat, which is used to produce steam to drive a turbine to generate electricity. The operation of a nuclear reactor and the related electricity-generating equipment is only one part of an interconnected set of activities.

The production of a reliable supply of electricity from nuclear fission requires mining, milling, and transporting uranium; enriching uranium (increasing the percentage of the uranium isotope U-235) and packing it in appropriate form; building and maintaining the reactor and associated generating equipment; and treating and disposing of spent fuel. These activities require extremely sophisticated and interactive industrial processes and many specialized skills.

Development

Britain took an early lead in developing nuclear power. By the mid-1950s, several nuclear reactors were producing electricity in that country. The first nuclear reactor to be connected to an electricity distribution network in the United States began operation in 1957 at Shippingport, Pennsylvania.

Six years later, the first order was placed for a commercial nuclear power plant to be built without a direct subsidy from the federal government. This order marked the beginning of an attempt to convert rapidly the world's electricity-generating systems from reliance on fossil fuels to reliance on nuclear energy. By 1970, 90 nuclear power plants were operating in 15 countries. In 1980, 253 nuclear power plants were operating in 22 countries, and by 2001 there were 435 nuclear plants operating in 33 countries. Despite this increase, the attempt to move from fossil fuels to nuclear energy faltered because of rapidly increasing costs, regulatory delays, declining demand for electricity, and a heightened concern for safety.

Safety Problems

Questions about the safety and economy of nuclear power created perhaps the most emotional battle fought over energy.

As the battle heated during the late 1970s, nuclear advocates argued that no realistic alternative existed to increased reliance on nuclear power. They recognized that some problems remain but maintained that solutions would be found.

Nuclear opponents, on the other hand, emphasized a number of unanswered questions about the environment: What are the effects of low-level radiation over long periods? What is the likelihood of a major accident at a nuclear power plant? What would be the consequences of such an accident? How can nuclear power's waste products, which will remain dangerous for centuries, be permanently isolated from the environment? These safety questions helped cause changes in specifications for and delays in the construction of nuclear power plants, driving up costs.

They also helped create a second controversy: Is electricity from nuclear power plants less costly, equally costly, or more costly than electricity from coal-fired plants? Despite rapidly escalating oil and gas prices in the late 1970s and early 1980s, these political and economic problems caused an effective moratorium in the United States on new orders for nuclear power plants.

This moratorium took effect even before the 1979 near meltdown (melting of the nuclear fuel rods) at the Three Mile Island nuclear power plant near Harrisburg, Pennsylvania, and the 1986 partial meltdown at the Chernobyl' plant north of Kyiv in Ukraine. The latter accident caused some fatalities and cases of radiation sickness, and it released a cloud of radioactivity that traveled widely across the northern hemisphere.

ENERGY EFFICIENCY IMPROVEMENTS

In addition to developing alternative sources of energy, energy supplies can be extended by the conservation (the planned management) of currently available resources. Three types of possible energy conservation practices may be described.

The first type is curtailment, that is, doing without closing factories to reduce the amount of power consumed or cutting back on travel to reduce the amount of gasoline burned. The second type is overhaul, that is, changing the way people live and the way goods and services are produced, using less energy-intensive materials in production processes, and decreasing the amount of energy consumed by certain products (such as automobiles).

The third type involves the more efficient use of energy, that is, adjusting to higher energy costs investing in cars that go farther per unit of fuel, capturing waste heat in factories and reusing it, and insulating houses. This third option requires less drastic changes in lifestyle, so governments and societies most commonly adopt it over the other two options.

By 1980 many people had come to recognize that increased energy efficiency could help the world energy balance in the short and middle term, and that productive conservation should be considered as no less an energy alternative than the energy sources themselves.

Substantial energy savings began to occur in the United States in the 1970s, when, the federal government imposed a nationwide automobile efficiency standard and offered tax deductions for insulating houses and installing solar energy panels. Substantial additional energy savings from conservation measures appear possible without dramatically affecting the way people live.

A number of obstacles stand in the way, however. One major roadblock to productive conservation is its highly fragmented and unglamorous character; it requires hundreds of millions of people to do mundane things such as turning off lights and keeping tires properly inflated. Another barrier has been the price of energy.

When adjusted for inflation, the cost of gasoline in the United States was lower in 1998 than it was in 1972. Low energy prices make it difficult to convince people to invest in energy efficiency.

From 1973 to the mid-1980s, when oil prices increased in the United States, energy consumption per person dropped about 14 percent, in large part due to conservation measures. Over time, improvements in energy efficiency more than pay for themselves. However, they require large capital investments, which are not attractive when energy prices are low.

Transportation

Whereas transportation uses 25 percent of the total energy consumed in the United States, it accounts for two-thirds of the oil used in the United States.

Cars built in other countries have long tended to be more efficient than American cars, partly because of the pressures of heavy taxes on gasoline. In 1975 the U.S. Congress passed a law that mandated doubling the fuel efficiency of new cars by 1985. This law, coupled with gasoline shortages in 1974 and 1979 and substantially higher gasoline prices (especially since 1979), caused the average efficiency of all U.S. cars to improve by about 40 percent between 1975 and 1990.

However, much of this improvement has been offset by dramatic increases in the number of cars on the road and by the growth in sales of sport utility vehicles and light trucks (which are not covered by federal efficiency standards). By 1999 the number of automobiles used worldwide had grown to 520 million vehicles, with an annual production of about 40 million automobiles.

This number is expected to increase to nearly 1 billion by 2018. Experts predict that unless more efficient technologies are developed, this growth will raise demand for gasoline by over 20 million barrels per day. Automobile manufacturers have the technical capability today to build cars with a much higher fuel efficiency than that mandated by Congress.

Mass-production of cars with this efficiency would require vast capital investments, however. New engine technologies that rely on electric batteries or highly efficient fuel cells, as well as engines that run on natural gas, may play a much greater role in the early 21st century. Increases in the prices of gasoline and parking have encouraged two other modes of transportation: ride sharing and public transportation.

Industry

Profit-conscious business managers increasingly

emphasize the modification of products and manufacturing processes in order to save energy. The industrial sector, in fact, has recorded more significant improvements in efficiency than either the residential or the transportation sector.

Improvements in manufacturing can be classified into three broad, somewhat overlapping, categories: improved housekeeping—doing routine maintenance on furnaces and using only necessary lighting; recovery of waste—recovering heat and recycling waste by-products; and technological innovation—redesigning products and processes to embody more efficient technologies.

Buildings

In early 1960s efficient energy use was often neglected in constructing buildings and houses, but the high energy prices of the 1970s changed that. Some office buildings built since 1980 use only a fifth of the energy used in buildings constructed just ten years earlier. Techniques to save energy include designing and siting buildings to use passive solar heat, using computers to monitor and regulate the use of electricity, and investing in more efficient lighting and in improved heating and cooling systems.

A life-cycle approach, which takes into account the total costs over the entire life of the building rather than merely the initial construction cost or sales price, is encouraging greater efficiency. Also, the retrofitting of old buildings, in which new components and equipment are used in existing structures, has been successful.

Chapter 6

Geothermal Energy

Geothermal energy is an important and promising alternative energy resource that has shown continual growth throughout this century; regrettably, its fortunes have reflected the variable successes experienced when traditional petroleum exploration techniques are used. Because the world's highest temperature—and perhaps most abundant—geothermal resources are associated with volcanic regions a framework for exploration and development of geothermal resources in volcanic areas by linking modern volcanological concepts to aspects of geothermal energy.

We emphasize the importance of volcanic field observations to geothermal exploration and review the OLADE approach to geothermal energy exploration. We have integrated quantitative approaches and models that can be used to collect and interpret field and laboratory data. These quantitative approaches have been introduced, in a simplified theoretical framework, to also show some links between volcanology and engineering concepts.

Volcanology has largely been an outgrowth of the larger discipline of geology and, like geology, is mostly a qualitative or "inexact" science. In contrast, much of the supporting science of geothermal energy has evolved from engineering methods that were developed in the petroleum industry; hence, it is intrinsically more quantitative and has a very different technical language.

This traditional dichotomy in technological approach has, we believe, hindered progress in both exploration for and developments of geothermal systems in volcanic areas. We practice a strategy that bridges that gap by first synthesizing classical and newly developed models of volcanoes (as well as their hydrothermal systems) and then applying this synthesis to the quantitative and engineering aspects of geothermal energy exploitation. This philosophy has been implicit in the OLADE methodology but has never been described and published in detail.

Because so much of volcanology has direct societal impact—and thus requires measures of certainty, the ability to predict, and inter-disciplinary approaches—volcanological studies have become increasingly quantitative over the last decade or more.

For economic minerals and geologic hazards applications, these studies must address such problems as investment security, environmental issues, and municipal safety. Our need for pragmatic approaches to research has spawned new methods in geophysics, mechanical engineering, and geochemistry; these techniques are now woven into the fabric of volcanology and have resulted in an increasingly quantitative discipline.

The inexactitude of the quantitative or semi-quantitative volcanologic models. Each volcano or volcanic area is complex and presents an individual problem to be solved systematically. Models only serve as a framework to focus and possibly enhance the efficiency of exploration and development. It is for this reason that we use case histories as examples for readers to consider in terms of typical models.

Finally, it is our intention that by addressing issues of volcanology that can be directly applied to engineering problems, it will be possible to better incorporate geologic reasoning into the development of geothermal resources.

Through our experience at Los Alamos National Laboratory, where geologists and engineers work closely, we have found that the "cross fertilization" of these disciplines produces seeds of understanding that can grow to strong exploration and development programs, often surmounting scientific and technological barriers that might have otherwise prevented success.

Application of Volcanological Observations to Geothermal Exploration. Our experience in numerous geothermal exploration projects has taught us a fundamental axiom for geothermal exploration in volcanic areas. Many of the complexities and unknown, subsurface characteristics of a volcanic geothermal field can be constrained through logical deduction that is based upon careful field observation, mapping, sample studies, and the integration of related geophysical and hydrogeochemical data.

We believe that many geothermal exploration projects in volcanic areas have suffered from the lack of pertinent volcanological observations and interpretations. So many clues regarding the location and magnitude of geothermal systems are available from the volcanic structure and deposits that one might say detailed interpretation of these observations constitutes a type of "exploration drillhole." Therefore, we stress the need for careful field volcanology during geothermal exploration projects in volcanic areas.

During the last 10 years, the field of volcanology has been growing rapidly; the resulting new observations and ideas are providing us with numerous hypotheses on volcanic structure and processes. In magma genesis, movement, and eruption phenomena, as well as volcanic structure and thermal histories, there have been many new discoveries that have engendered a better understanding of igneous systems and their relationship to high-grade geothermal systems. These hypotheses and discoveries have important geothermal implications when applied to the interpretation of volcanological observations.

Geologists must use what is known about volcanoes, their structure, eruption phenomena, and composition, to reveal necessary information about the heat sources and settings of groundwater—key factors in formation of a hydrothermal system. A basic approach to exploration includes good geological mapping by whatever means is available: topographic maps, aerial photographs, satellite photographs, planetable surveying, tape and brunton traverses, and panoramic viewpoints. Also, systematic descriptions of tephra deposits and rocks are vital, especially for core logs from exploration holes.

In applying volcanological observations, one should integrate the observations (for example, mapping and sample analyses) with other information on surface springs and fumaroles, water chemistry and hydrology, and geophysical surveys, including gravity, electrical resistivity, seismicity, and heat flow. Any of these surveys by itself, without a geologic framework, is almost useless; integrated with good geological surveys, each is valuable. Hydrochemists, geophysicists, reservoir engineers, and geologists must talk to each other and work as teams to successfully develop geothermal resources.

A basic methodology for geothermal exploration in volcanic fields was developed in 1983 by an international team of experts for the Latin American Energy Development Organization (OLADE, 1983). The field approach involves learning everything possible about a volcano or volcanic field, including structure, structural setting, eruption phenomena, composition, and ages of eruptions.

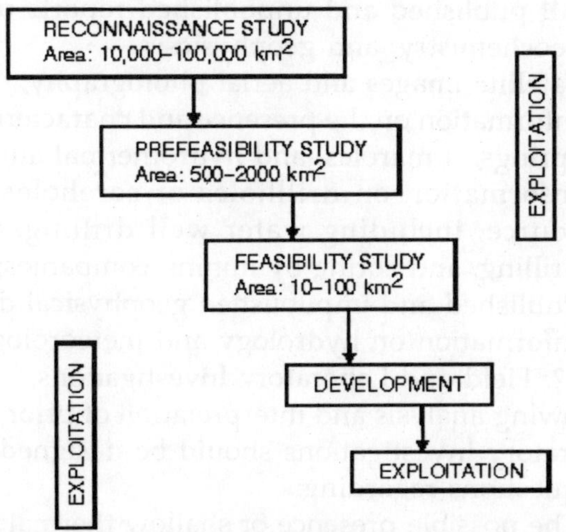

Fig. Flow Diagram Showing Steps of the OLADE Methodology for a Geothermal Project.

Using these data, it is possible to establish hypotheses regarding the location and magnitude of hydrothermal resources. Simply put, the volcano and its products supply information normally gathered from the first drillholes and may provide a view of the volcano's geothermal system or systems. Employed in conjunction with hydrogeochemical sampling of hydrothermal waters and gases,the field approach is extremely cost-effective and is the first step toward prescribing the more expensive geophysical surveys and exploration drilling that will be needed.

The geological and physical volcanological approach involves the three steps outlined here; these steps are undertaken at the same time as the hydrogeochemical sampling but before geophysical surveys.

Step 1: Evaluation of Available Information about the Area to Be Studied. In the initial phase of a project, all existing information is collected and evaluated, including:

- Topographic and geological maps at large and small scales;
- Regional geological syntheses, including stratigraphy, structural geology, and history of volcanism;
- All published and unpublished reports on geology, geochemistry, and geophysics;
- Satellite images and aerial photography;
- Information on the presence and characteristics of hot springs, fumaroles, and hydrothermal alteration;
- Information on drillholes or coreholes from any source, including water well drilling, petroleum drilling, and coring by mining companies;
- Published and unpublished geophysical data; and
- Information on hydrology and meteorology.

Step 2: Field and Laboratory Investigations.

Following analysis and interpretation of prior work, field and laboratory investigations should be designed to answer specific questions regarding:

- The possible presence of shallow thermal anomalies,
- Regional hydrologic conditions, and
- The nature of thermal manifestations. In volcanic regions, it is important to focus geological observations on a number of points.

- Identify those areas where there are episodes of recent volcanism. The definition of "recent" varies according to the volume of material erupted because large magma bodies retain heat much longer than small ones do.
- Evaluate the relative quantities of silicic and mafic or intermediate volcanic products.
- Define, on a regional scale, the present relationship between the volcanic structure and the regional tectonic framework.
- Identify phreatic explosion craters.
- Systematically collect samples of all lithologic types for laboratory analysis, including petrographic and chemical analyses.
- Collect lithic clasts (xenoliths) from pyroclastic units for petrographic analysis.
- Determine the absolute ages of representative lithologic units.
- Study (in preliminary form) all possible reservoir and caprock units.

Analysis and interpretation of field and laboratory data at this time will help define principal geothermal areas to be studied in detail and, if appropriate, selected for geophysical surveys and exploratory drilling. Along with results of the regional hydrogeochemical surveys, the preliminary data can be used to determine areas to be evaluated for potential commercial development.

Step 3: Detailed Field and Laboratory Studies: Geology and Volcanology

Detailed field and laboratory studies begin with:

- Interpretation of aerial photography,
- Preliminary identification of faults and volcanic structures,
- Hypotheses concerning the regional volcanotectonic setting, and
- Integration of information from existing maps.

Following this work is a detailed field study that comprises the aspects listed here.

- A search for thermal anomalies in the upper crust involves mapping and sampling young volcanic eruption sequences, especially rock types indicative of shallow magma bodies. All areas of hydrothermal manifestations, both fossil and active, are mapped and sampled in conjunction with hydrogeochemical sampling. All volcanic structures are mapped, including craters, domes, phreatic craters, and associated faults.
- In areas with surface hydrothermal manifestations, potential caprocks are mapped and sampled, and their origin is determined. In volcanic zones, the search for phreatic explosion craters is emphasized.
- The extent of potential geothermal reservoirs can be estimated through
- A study of lithic clasts (xenoliths) in pyroclastic deposits; these clasts provide information on the nature of rock units underlying the volcano.
- Identification and mapping of recent faults. This effort is essential because active faults frequently represent zones of fracture permeability.
- Determination of the degree of hydrovolcanic activity responsible for pyroclastic deposits in the volcanic field. This work may identify aquifers beneath the volcano during recent eruptions. These aquifers could be current hydrothermal reservoirs.

In tropical countries where soils form rapidly and outcrops are soon covered by vegetation, geological mapping is considerably more difficult.

In these situations, several additional approaches are necessary:

- Landform mapping. These maps are based primarily on the interpretation of aerial photographs and satellite images, especially in young volcanic fields. The interpretations are field checked along road cuts, stream bottoms, and shorelines, as well as in quarries.
- Side-looking airborne radar (SLAR) imagery. Such images are extremely useful in mapping faults and

volcanic landforms in tropical areas, although they may be relatively expensive to acquire.

Recent Practical Advances in Volcanology

Quantitative methods for studying volcanoes and their products are gaining importance in the evolving field of volcanology. Using increasingly more precise and accessible laboratory techniques to determine chemical compositions of rocks and minerals, petrologists have developed methodologies to understand the origins and evolution of magma.

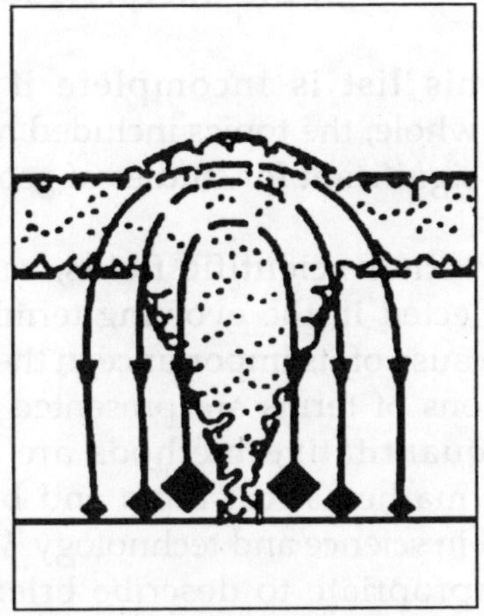

Fig. Quantitative Volcanoes

The constraints on temperature and pressure estimated from chemical data are enhanced by results of geophysical surveys; together, these efforts have led to a better understanding of magma-chamber dimensions and locations. Simultaneously, the development of computer capabilities has allowed volcanologists to systematically quantify field observations that can be numerically modeled by using fluid mechanics.

Although these developments are still relatively immature compared to similar work in other scientific fields, the advances constitute a tremendous resource for practical application in geothermal energy exploration.

Advances in volcanology that we consider to be fundamentally important for understanding geothermal resources:

- Documentation and interpretation of chemical zonation in large-volume pyroclastic deposits and consequential implications concerning the nature of magma chambers,
- Development of quantitative approaches for describing and understanding eruption dynamics and the emplacement of volcanic products, and
- The hydromagmatic theory of eruption mechanics and its significance in the interpretation of pyroclastic deposits.

Although this list is incomplete if one considers volcanology as a whole, the topics included here are those we have found to be significant for studies of geothermal systems in volcanic areas.

As in many other scientific fields, rapidly changing technology is reflected in the evolving terminology, and we stress this fact because of its importance in the communication of ideas. Definitions of terms are presented in the Glossary. Terms used in quantitative methods are especially vital because of their mathematical roots and because they are widely employed in science and technology. With this concept in mind, it is appropriate to describe briefly the common quantitative approaches taken in volcanology.

Quantitative Methodology and Volcanology

The consequence of maps as the fundamental method of data representation is perhaps unique to the science of geology. Volcanology certainly relies heavily upon maps to graphically portray research problems, their geographic locations, physical and chemical trends, and hypothetical arguments. On a map, many types of quantitative data can be portrayed.

For example, mapping contours of deposit thickness (*isopachs*) or clast diameters (*isopleths*) within the deposit is a common technique in tephra deposit studies. The mathematical representation of these contours is a valuable method for locating vent areas and estimating the volume of eruptive products as well as their emplacement mechanism.

The first step in analyzing a map-oriented data set—especially if a hypothetical argument is lacking—is statistical analysis of data trends. Examples of geological data trends are discussed extensively by Davis (1973), and specific cases include:

- Trend-surface analysis of stratigraphic units and topography,
- Rose diagrams of structural fabrics,
- Geographic correlation of absolute and relative rock ages,
- Areal density of specific surface features, and
- Cluster analysis of geochemical data to define major variations.

Following statistical analysis, hypothesis development and testing can be undertaken, and these generally focus upon chemical and physical problems. Data correlation makes it possible to assess underlying physical or chemical controls in cases where independent and dependent variables are hypothesized. Least squares, regression, and Fourier techniques are commonly applied to correlated data. The use of multivariate analysis is an issue of greater controversy, but the method may have significant application in classification schemes. The following pages outline some fundamental physical and chemical relationships that directly apply to geothermal systems.

Physical Processes

Energy transfer through the earth's crust (and heat flow in particular) is a basic component of geothermal systems and volcanoes. In general, heat flow is influenced by several processes that sum in the following equation (Shimazu, 1963).

$$pC\frac{dT}{dt} = \alpha T\frac{dp}{dt} + \frac{1}{r^2}\frac{\partial}{\partial r}k_t r^2\left(\frac{\partial T}{\partial r}\right) + \Sigma H_i \exp(-\lambda_i t) + \Sigma J_i$$

where temporal heat flow (r = density, C = specific heat, T = temperature, and t = time) is equal to the sum of the adiabatic temperature gradient (a = the coefficient of thermal expansion, p = pressure), heat conduction (r = radial distance, k_t = the thermal conductivity), radioactive decay heat (H_i = heat liberated by decay of the i th isotope, l_i = decay constant), and heat of reaction (J_i = the heat produced or liberated for the ith chemical reaction).

To estimate heat flow for volcanic and geothermal systems, this expression must be altered to include convective heat flow, which can be approximated by replacing in the above equation several elements:

$$\frac{dT}{dt} = \frac{\partial T}{\partial t} + u_{\text{conv}} \frac{\partial T}{\partial r}$$

where u_{conv} is the velocity of convection, which can in turn be approximated by a function of Rayleigh number: u_{conv}@ $3(Ra)^{1/3}$, where $Ra = (ra\ gD\ TD^3)/(d_t\ \mu)$, and μ = viscosity, g = gravitational acceleration, D = a characteristic length of the flow, and $d_t = k_t/(rC)$.

Studies of mass transfer associated with volcanism generally focus on movement of magma and magmatic volatiles from the magma chamber to the surface of the earth.

Two extreme cases of these processes are:

- Eruptions that result in effusions of lava, and
- Explosive eruptions in which the expansion of gases determines mass transfer processes. Incompressible approximations of mass and momentum conservation are useful descriptive equations. For passive magma flow in conduits, the Bernoulli equation is

$$p_1 = p_2 + p\left(\frac{u_2^2 - u_1^2}{2}\right) + pg(h_2 - h_1) + pf_h$$

where subscripts denote values measured at two different levels in the conduit or flow system, h_1, h_2, u = velocity, and f_h = a term reflecting frictional losses and is a function of conduit or substrate surface roughness and Reynolds number:

Re = $(r\ uD)/\mu$. Where the expansion of volatile phases under conditions of changing pressure and temperature causes both fragmentation of magma into tephra and rapid acceleration of a gas and solid mixture from the vent, it is possible to write the Bernoulli equation to account for changes in gas pressure.

Two end-member processes of gas decompression are

- The adiabatic case in which no heat is exchanged between the gas and solid particles and
- The isothermal case, in which heat is continuously supplied to the gas from the tephra during decompression. Both cases depend upon the gas weight fraction in the mixture:

$$n = \left[\frac{p_p p_g}{p_b} - p_g\right] / (p_p - p_g)$$

where r_g = the gas density, r_b = the bulk density of the mixture, and r_p = the solid particle density. The adiabatic and isothermal cases are respectively:

$$\text{nRT}\frac{\gamma}{\gamma-1}\left[1-\left(\frac{p_2}{p_1}\right)^{(\gamma-1)/\gamma}\right]+\frac{1-n}{p_p}(p_1-p_2)$$
$$= g(h_1-h_2)+\left(\frac{u_2^2-u_1^2}{2}\right)+f_h$$

and

$$\text{nRT}\left[\ln\left(\frac{p_2}{p_1}\right)\right]+\frac{(1-n)}{p_p}(p_1-p_2)=g(h_2-h_1)+\left(\frac{u_2^2-u_1^2}{2}\right)+f_h$$

where g = the gas isentropic exponent.

Various adaptations of Eqs. can be applied to different eruptive conditions they are then useful in approximating the basic relationships among gas pressure, temperature, and abundance as well as exit conditions such as ejecta velocity and column height. However, these approximations and calculations yield accurate results only when considered in light of nonlinear relationships like those included in the full set of Navier-Stokes equations written separately for gas and solid phases.

Chemical Processes

During the past decade, volcanic petrologists have made

great strides in understanding the complex origins of magma chemistry as revealed by analyses of phenocryst and glassy components of volcanic products. Because these analyses provide abundant quantitative data, mathematical approaches are particularly suited for modeling the origins of chemical signatures. Magma composition generally evolves with time as a result of:

- Initial melting from source rocks,
- Fractional crystallization caused by cooling and the loss of volatile constituents, and
- Comingling with magmas of different composition (Carmichael *et al.*, 1974). The behaviour of chemical species during these three important differentiation processes can be quantitatively modeled by using chemical data provided through bulk and modal analyses in which trace-element behaviour is most indicative of the differentiation mechanism.

The Rayleigh equation applies to fractional crystallization and predicts the concentration of a particular chemical species remaining in the liquid (c_1) after crystallization of a specific fraction of crystalline phase when the original species concentration is c_o:

$$c_1 = c_o F^{(k_{o-1})}$$

k_d = the Nernst distribution coefficient, which expresses the fraction of the chemical species in the liquid that enters the crystalline phase. F = the fraction of original melt remaining. For cases in which the chemical species enters two or more phases, k_d is replaced by D_o, the weighted average of solid-liquid partition coefficients of all the phases. If crystallization is incomplete at some value of F, then c_1 given by Eq. must be multiplied by k_d or D_o, depending upon the number of phases involved.

For conditions of partial melting, in which the liquid phase remains in equilibrium with the residual solid phases until it is removed, the Berthelot-Nernst equation predicts c_1 by

$$c_1 = \frac{c_o}{D_o + F(1-P)}$$

Here P = the bulk partition coefficient for the phases that melt, and F = the fraction of melted material. When only one phase is melted, $P = D_o$.

Fractional crystallization and partial melting result in a logarithmic relationship between species concentration in the solid/liquid phases and the degree of melting or crystallization.

When both fractional crystallization and partial melting occur and D_o approaches zero, Eqs. (1-7) and (1-8) reduce to $c_1 = c_o/F$. If one assumes that fractional crystallization occurs when there is equilibrium between the total crystallizing solid and melt, its description takes a form analogous to that of partial melting [Eq. (1-8)].

For situations in which chemical trends are the result of mixing two magmas of different compositions, a mass balance equation (Gast, 1968) predicts the resulting concentration in the magma (c_x) of some species; m_{m1}, c_1, m_{m2}, and c_2 are the magma mass and species concentration of magma 1 and magma 2, respectively:

$$c_x = \frac{m_{m_1} c_1 + m_{m_2} c_2}{m_{m_1} + m_{m_2}}$$

In contrast to fractional crystallization and partial melting, the concentration of a chemical species in a mixed magma is linearly dependent on the degree of mixing.

Isotopic tracers are also very useful for determining the origin and evolution of magmas. As in the case of stable isotopes such as oxygen, the isotopic composition is related to a standard. For oxygen, the heavy-isotope ^{18}O abundance is expressed

$$\delta^{18}O = \left(\frac{(^{18}O/^{16}O)_s - (^{18}O/^{16}O)_{smow}}{^{18}O/^{16}O)_{smow}} \right) \bullet 1000\%,$$

where the subscripts s and smow denote the sample and standard mean ocean water isotopic ratios, respectively. Whereas stable isotopes are considered excellent chemical tracers, radiogenic isotopes are employed in dating techniques and are widely used in geochemistry.

Magma Generation, Accumulation and Differentiation in Chambers, and Eruptions. Recent geochemical studies in igneous petrology have focused on the processes of magma generation, evolution, and collection in subsurface reservoirs called magma chambers. Although chemical reactions continue to change the composition of lava and tephra after these materials reach the earth's surface, petrologists traditionally studied only the history of volcanic rocks before their eruption.

Major fields of interest have been the tectonic setting and origin of magma, processes of chemical differentiation, and magma-chamber dynamics. Recently, with the advent of powerful computers, heat flow, seismicity, and fluid convective and diffusive processes have been used to develop comprehensive models of magma chambers. Through geochemical analysis, field samples of volcanic products yield vital clues about the parent materials of magmas, the depth of their generation, and the differentiation processes that affected them on their path through the earth's crust.

Tectonic Setting and Origin of Magmas

Because ~95% of all volcanoes occur at plate margins, their locations are consistent with the theory of plate tectonics. The magma sources of volcanoes that occur in intraplate areas, whether oceanic or continental, are more difficult to explain. In these cases, hypotheses that involve mantle

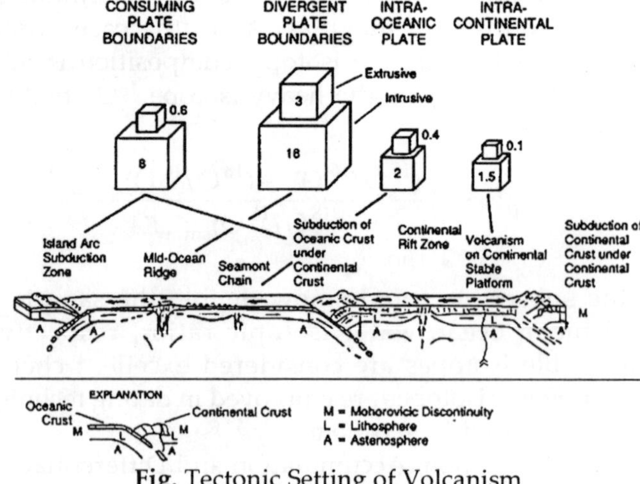

Fig. Tectonic Setting of Volcanism

dynamics, such as hot plumes associated with mantle convection cells continental rifting and lithospheric thinning associated with extensional tectonics appear to be reasonable. Perhaps the most significant aspect of a tectonic setting is its effect on observed magma compositions and chamber development.. A primary characteristic for geothermal potential is the chamber depth, which provides information about the magma source and stagnation depth (the geothermal heat source depth).

For mantle-derived magmas, which are mafic and bear mantle signatures of trace elements and isotopic ratios, source depths of > 50 km are expected and—in the cases of continental, intraplate volcanoes—may show no crustal reservoirs. On the other hand, in rifts and extensional terrains, deep mantle magmas promote melting of crustal rocks so that shallow silicic magma chambers can develop during long periods of magma flux from the mantle. In contrast, arc-related volcanoes show the effect of crustal thickness. Continental arcs have magmas that, having been generated at intermediate depths of several tens of kilometers, may stagnate or become contaminated by more siliceous crustal materials during their assent.

Magma Chambers

Volcanic products are generally classified by their major-element chemistry or their modal phenocryst content. These classification schemes are useful in relating volcanic rocks to magma types. Accordingly, the origin and evolution of magma types can be interpreted in a general manner by considering igneous compositional trends: tholeiitic, transitional, alkalic, potassic, and calcalkalic.

When the field geologist examines pyroclastic samples that do not lend themselves to the above classification schemes, the colour of glass shards can be simply related to their refractive index as a function of silica content. Rock classification has been a traditional exercise for volcanologists, and today the results of this work can be used to determine the nature of the magma source: its shape, depth, and longevity—all of which are important components when evaluating geothermal potential. Nonbasaltic volcanic rocks are considered to be products of evolved magmas. Hildreth stated, "every large eruption of nonbasaltic magma taps a magma reservoir that is thermally and compositionally zoned," and "most small eruptions also tap parts of heterogeneous and evolving magmatic systems."

One general hypothesis is that evolved or otherwise differentiated magmas have a crustal reservoir. The volume of a crustal magma chamber is directly proportional to the time required for it to evolve. Consequently,differentiated volcanic products—especially where they are several cubic kilometers in volume—are good indicators of a crustal magma chamber.

Table: Characteristics of Volcanic Systems and Tectonic Settings

Tectonic Setting	*Regional Stress*	*Magma Volume*	*Dominant Composition*	*Chamber Depth*
Island Arcs			Mafic-Intermediate	Intermediate
Continental Arcs		25.6	Intermediate-Silicic	Shallow
Rifts	Extensive	62.5	Mafic-Silicic	Deep-Shallow
Intraplate	Various	5.5	Mafic	Deep-Shallow

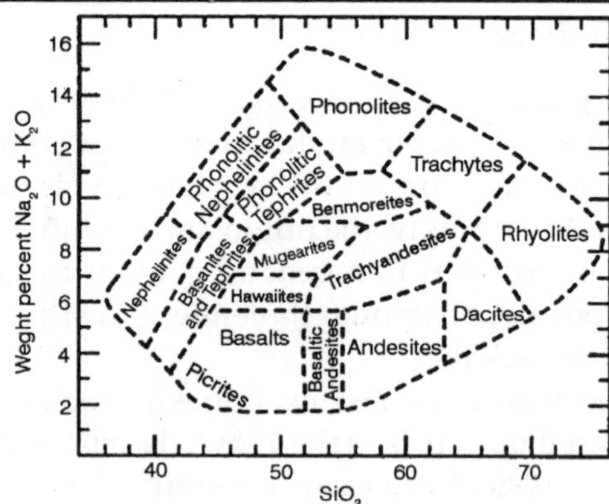

Fig. Classification of Volcanic Rocks

Observation of the correlation between caldera area and ejecta volume opened the door for interpretation of chemical zonation in silicic magma chambers.

He predicted that "all caldera-forming ash-flow sheets

should, when studied in detail, show some degree of chemical and/or mineralogic gradients inherited from the magma chamber." Hildreth documented such gradients in the Bishop Tuff in eastern California.

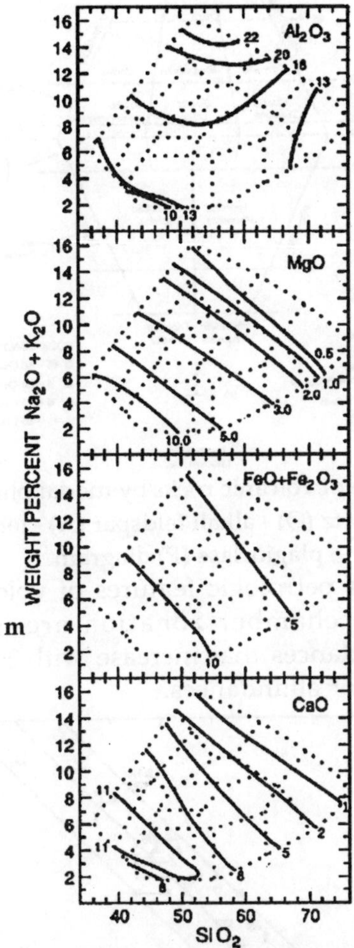

Assuming that earlier erupted products originate from the top of a magma chamber and later materials derive from lower portions, it is likely that the time-sequenced chemical characteristics of volcanic ejecta depict an inverse order of the magma chamber's compositional stratification. This chemical stratification also is reflected by oxygen fugacity and mineral equilibrium temperatures that increase with time in products from a large eruption.

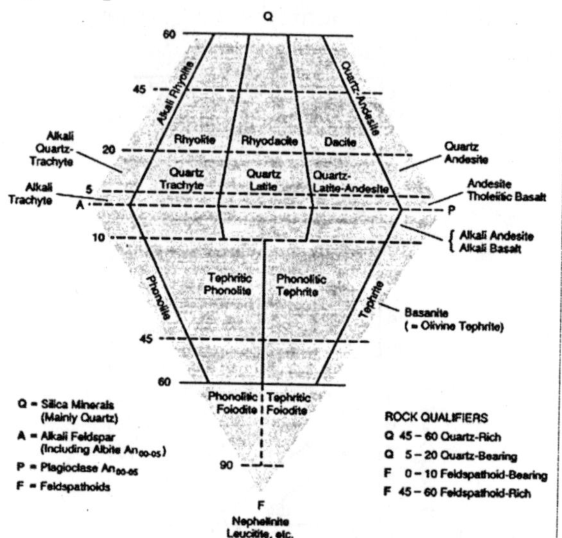

Fig. Classification of volcanic rocks by modal phenocryst content plotted on a quartz (Q) - alkali feldspar (A) - feldspathoid (F) - plagioclase (P) diagram.

Several other petrologic features of volcanic ejecta that suggest magma chamber zonation are isotopic ratios, phenocryst abundances that increase with SiO_2 values, and volatile component abundances.

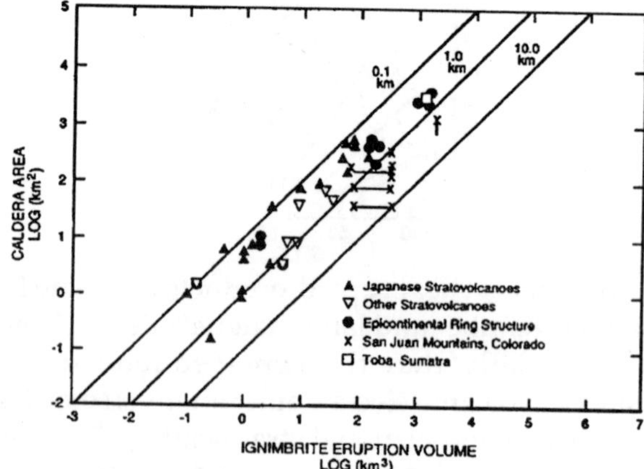

Fig. Correlation Between Caldera Area and Volume of Products

This latter feature is best typified by stratigraphic relationships showing that early products resulted from more explosive, gas-rich eruptions and later materials were from gas-poor effusive extrusion. Volatile zonation in rhyolitic eruptions might only reflect eruptive conditions. In Eichelberger's model, the volatiles are not stratified in the magma. Early eruptions are explosive because the volatile flux is confined within a narrow vent region, whereas later effusive eruptions involve a gradual degassing of the rhyolite through permeable vent-wall rocks—a process that results in a volatile-poor magma by the time it reaches the surface and is extruded as a lava flow.

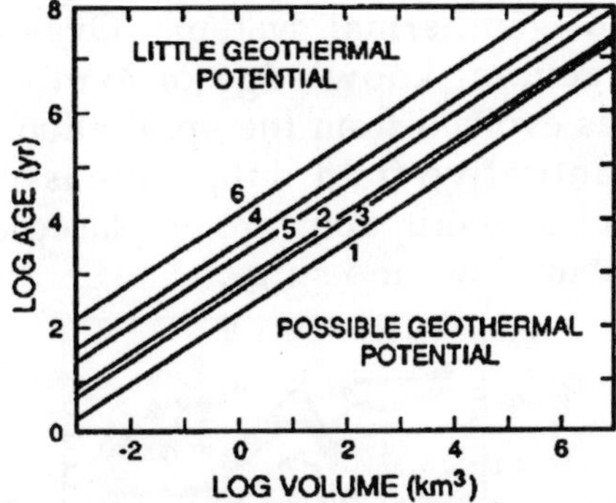

Although major-element abundances do support hypotheses of magma-chamber zonation, it is analyses of the trace elements that best portray the nature of the zonation and mechanisms of differentiation as a result of their variable compatibility in various phenocryst and liquid phases.

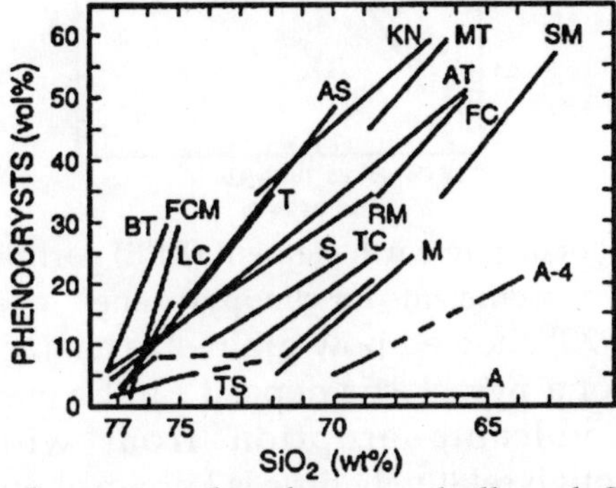

Fig. Plot of phenocryst abundances vs bulk-rock SiO_2 of silicic magma for various volcanoes.

Petrologic studies of magma suggest that large chambers are fundamentally basaltic because mantle melting supplies heat to the crust for crustal melting, provides a mafic component to hybridize with the crustal melts, and generates a thermal gradient to drive various differentiation processes in the crustal magma reservoir. This general evolution of crustal magma chambers may depend upon tectonic environment.

Explosive Eruptions and Quantitative Models

Explosive volcanic eruptions are significant in the development of geothermal systems. Over the past two decades, our general knowledge of explosive eruption mechanisms has evolved from the application of theoretical models to quantitative field data. The assumption that products of explosive eruptions are emplaced as pyroclastic deposits by fallout, flow, and surge.

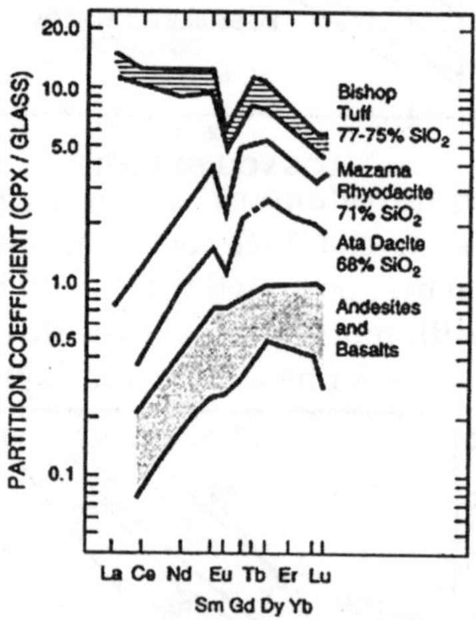

Fig. Plot of rare-earth element (REE) partition coefficients for clinopyroxene

Walker (1973) showed how grain-size characteristics and dispersal area of a pyroclastic deposit can be used to deduce the type of volcanic eruption from which it was produced.Although relatively little is known about subsurface processes in the volcanic conduit, the behaviour of eruption columns has been deduced from observations; this information allowed Wilson and Sparks and Wilson to formulate physical conditions in explosive eruption columns.

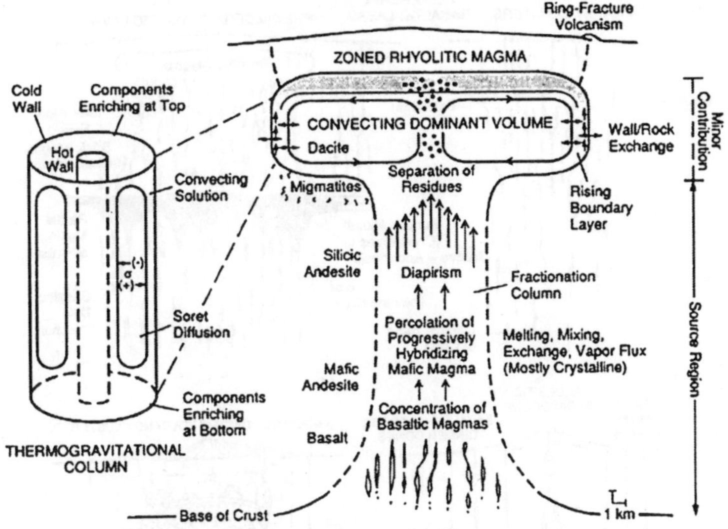

Fig. 1.9Processes affecting magma chamber

The basic equations for the eruption are

$$\frac{dp}{p} + \frac{du}{u} + 2\frac{dr_c}{r_c} = 0$$

and

$$-\frac{dp}{p} = gdh + udu + \frac{f_h u^2}{4r} dh$$

which express conservation of mass and momentum, respectively, for one-dimensional flow along the subsurface volcanic conduit.

The h and r_c = vertical distance and conduit radius, respectively; g = gravitational acceleration, u = the magma's velocity, $p = r_g$RT (perfect gas law pressure), and r = bulk density. f_h is the factor expressing frictional loses along the conduit walls. The relationship among bulk (r_b), solid (r_p), and gas (r_g) densities is expressed as in Eq.

This quantitative approach to understanding volcanic phenomena is well summarized by Head and Wilson for a variety of eruption types, including effusive processes, Strombolian (scoria cone), Hawaiian (lava fountain), Plinian

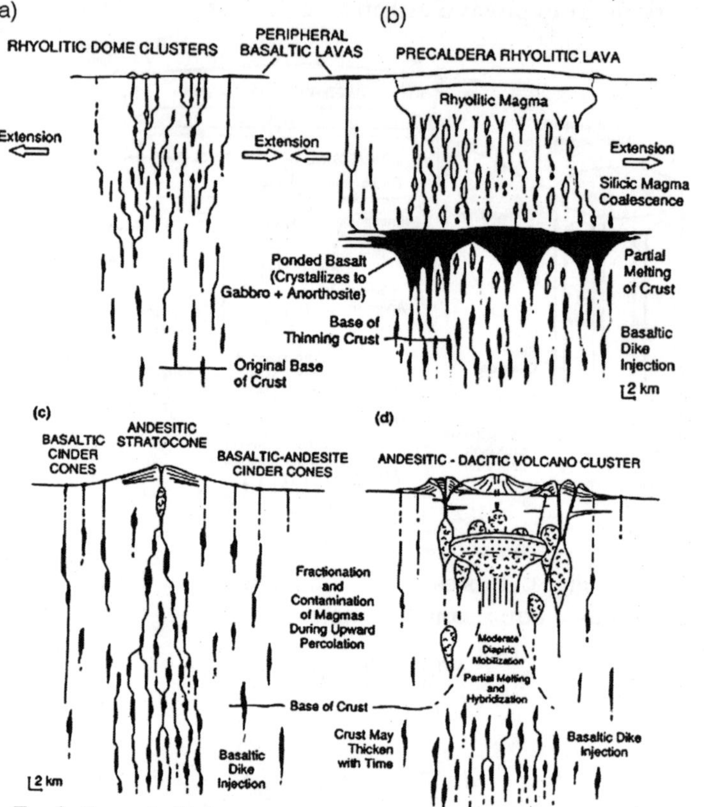

Fig. Evolution of silicic magma chambers as a function of tectonic environment. (a) early and (b) advanced stages (c) early stage and (d) intermediate stage.

Pyroclastic Fallout

Pyroclastic fall deposits are characterized by their relatively well sorted size characteristics, topography-blanketing dispersal, and graded bedding, but lack of other internal bedforms. The emplacement characteristics of these deposits are controlled by the terminal fall velocities of individual pyroclasts. One important component for this modeling is the assumption that eruption columns behave as thermal plumes in which the height of the plume (h_t) is proportional to the quarter root of the mass flux (d m/d t):

$$h_t = k_h(dm/dt)^{1/4},$$

The constant of proportionality (k_h) is ~43.7 for steady columns and 7.22 for discrete explosions when d m/d t is expressed in kilograms per second (1 kg/s @ 1.1 kW) and h_t in meters. For a convecting eruption column, a second important assumption is that vertical velocities (u_v) fit a gaussian function of distance from the plume axis:

$$u_v = u_c \exp(-x_r{}^2/b_e{}^2),$$

where u_c = the centerline velocity at height h as determined from solutions of Eqs. x = the radial distance from the plume axis, and b_e = the e-folding distance of u_c; $2b_e$ is the approximate distance from the plume axis to the visible edge of the plume. Superimposed upon u_v is u_r, the radial velocity of lateral plume spread, which is defined as

$$u_r = \frac{(dm/dt)}{2\pi r_p p_a (h_t - h_b)}$$

where r_p = the plume radius, r_a = the mean air density between h_t (the plume height) and h_b (the height at which the plume is neutrally buoyant and begins appreciable lateral movement).

Pyroclastic Flows

Pyroclastic flows (ignimbrites) comprise some of the most voluminous explosive products in the geologic record, and one possible emplacement model is that for the gravitational collapse of an eruptive column. Based upon Prandtl's theory of turbulent fluid jets, in which ambient air is incorporated into the jet—thus changing its bulk density, the equation of motion for an eruptive column:

$$-\frac{g}{q^2}\left(1-\frac{p_a}{p_b}\right) = u\left(\frac{du}{dh}\right) + \frac{(u^2/2+gh)}{r_v^2 b_p}\frac{d(r_v^2 p_b)}{dh}$$

where q = a ratio of the average column velocity to its centerline velocity, r_b = the bulk density of the column, r_v = the vent radius, and r_a = the density of the ambient air. Numerical solutions to this equation, relate column height to gas velocity, vent radius, and water content. Column collapse is predicted for columns that do not continue their upward motion because buoyancy forces can no longer offset drag forces on the margins of the column.

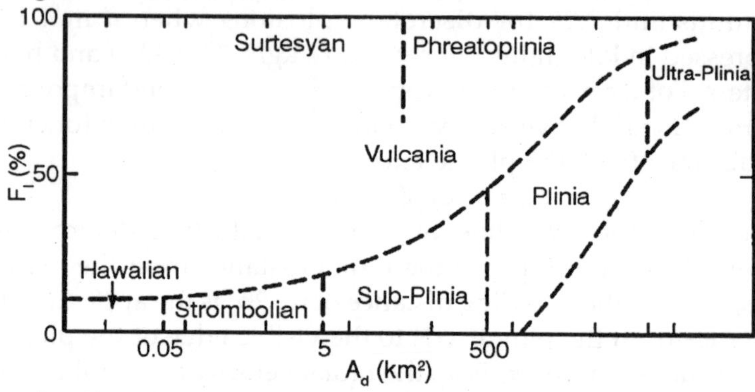

Fig. Classification of Eruptive Mechanism by Grain Size and Dispersal

The onset of gravitational collapse predicted by solutions to Eq. Plinian eruptive column collapse can be precipitated by increases in vent radius or decreases in the water content of erupting materials; either condition decreases the initial velocities of the column and leads to its collapse.

Modeled the runout of pyroclastic flows and surges by employing an "energy line" concept derived by analogy to rock-fall debris streams, which are dominantly gravity-driven flows. The maximum distance of runout is computed as the loci of points at which the potential energy surface of the flow intersects the topographic surface. The velocity of the flow at any increment along its flow path is simply modeled as its gravitational potential velocity path: $v(i) = [2gD\, h(i)]^{1/2}$, where $D\, h(i)$ = height of the energy surface above the local topography; in general, this value is initially determined by height above the vent from which the pyroclastic flow collapses. For directed blasts the initial velocity can be taken as a calculated gas-dynamic velocity such as the blast's sound speed. The flow accelerates with incremental runout distance:

$$a(i) = g[\sin\theta(i) \cdot \mu h \cos\theta(i)]$$

for which q (i) = the local slope and μ_h = the tangent of the energy surface slope (q_e), called the Heim coefficient. This number can vary from 0.06, for highly mobile, large pyroclastic flows, to 0.74, for small pyroclastic flows with low mobility (Sheridan, 1979).

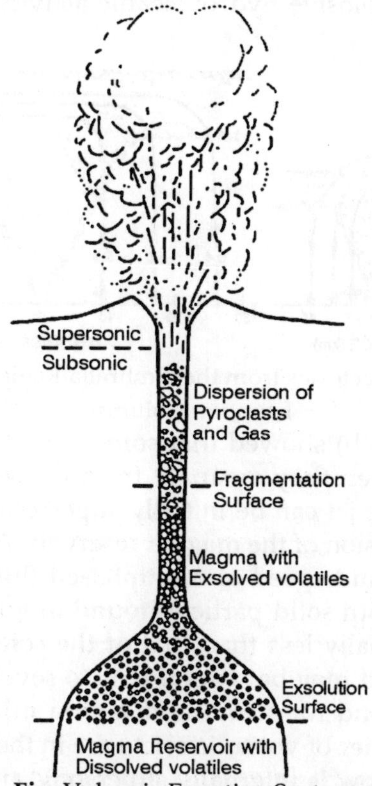

Fig. Volcanic Eruptive System.

The flow accelerates and decelerates depending upon the local slope, in such a way that it flows over a total runout distance (L_f) to where its velocity v(i) = 0; $v(i) = [v_o + 2a(i)L_e(i)]^{1/2}$, where L_e (i) is measured from topographic maps and $t(i) = 2L_e(i)/v(i)]$.

Pyroclastic Surge

Relatively thin bedding, and a multiplicity of bedforms distinguish the deposits of pyroclastic surges. These textural features are thought to indicate unsteady flow and rapid variations in particle-to-gas volume ratios—flow conditions that are especially prevalent during eruptive blasts such as those that may occur during the initial moments of Plinian eruption and explosive hydrovolcanic activity.

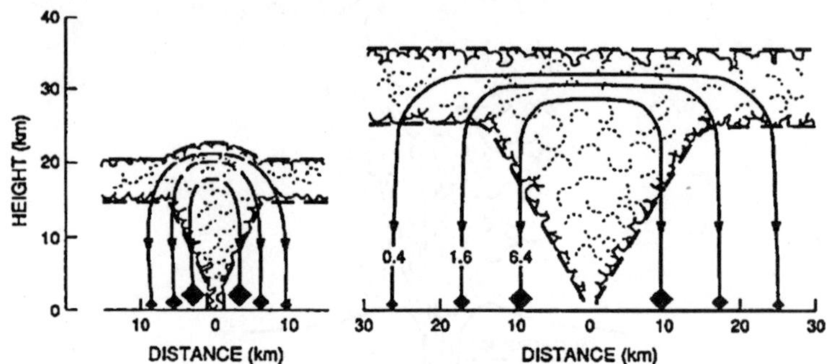

Fig. Clast Trajectories from the Umbrella Region of a Plinian Eruption Column

Kieffer (1984b) showed that some volcanic blasts have a jet structure when they emanate from the vent orifice. The conditions of the jet can be initially supersonic and will vary with decompression of the magma reservoir. As Kieffer (1977) showed, the sound speed of multiphased fluids (c_s), such as steam loaded with solid particles found in volcanic columns, can be substantially less than that of the constituent phases. The sound speed may be several tens to several hundreds of meters per second for steam and tephra mixtures. Because observed velocities of volcanic ejecta are in the range of 100 to 500 m/s, their flow is internally supersonic and the effects of gas compressibility are important. The Bernoulli Eq. can be written to show the effect of Mach number ($M = u/c_s$):

$$p_0 = p_s\left(1+\frac{\gamma+1}{2}M^2\right)^{\gamma/(\gamma-1)}$$

in which p_o = the stagnation pressure (the pressure of the erupting mixture at zero velocity; for example, the chamber overpressure), p_s = the static pressure, and g, the isentropic exponent (ratio of heat capacities at constant pressure and constant volume), expresses the degree to which the erupting mixture approaches isothermal expansion (g = 1.0). In contrast to the incompressible Bernoulli Eq. in which the pressure is a function of velocity only, the compressible form shows that pressure is also a function of thermodynamic parameters. For eruption columns modeled by incompressible equations, the pressure along the axis of the column is nearly atmospheric, but for columns erupted as supersonic jets, the effects of compressibility cause pressure and density to vary by large factors along the column's axis.

To understand flow conditions for surge-producing blasts, it is necessary to solve non-linear forms of the equations of motion. In simplified form, these equations express

$$\vec{\nabla} \bullet (p\vec{u}) = 0 \qquad \text{(Continuity)}$$

$$\vec{\nabla} \times \vec{u} = 0 \qquad \text{(Irrotationality)}$$

and

$$\vec{\nabla}\left(\frac{u^2}{2}\right) + \frac{1}{p}\vec{\nabla}p = 0 \qquad \text{(Momentum)}$$

where r = density, $\vec{u}$ = the velocity vector, and $\vec{\nabla}$ = the nabla operator that signifies spatial differentiation. Unlike previous models of eruption columns, these equations cannot be solved analytically, which is the main reason previous researchers used incompressible approximations. However, using the classical method of characteristics, Kieffer obtained solutions for the continuous ranges of the equations to show their profound effect upon the flow of tephra and gas during blast eruptions.

A more complete formulation of this problem involves the complete set of multiphase, Navier-Stokes equations and employs a high-speed computer. However, the emplacement of pyroclastic surges, a topic of great importance in volcanic hazard analysis, has not been so completely analyzed that quantitative models can predict field relationships.

The important quantitative models includes those that have had wide applications in recent years and are frequently cited. With improved modeling approaches and close development of theory in conjunction with field observation, it will be possible to use field measurements to constrain eruptive mechanisms and subsurface conditions that are needed to understand the thermal regime and hydrothermal

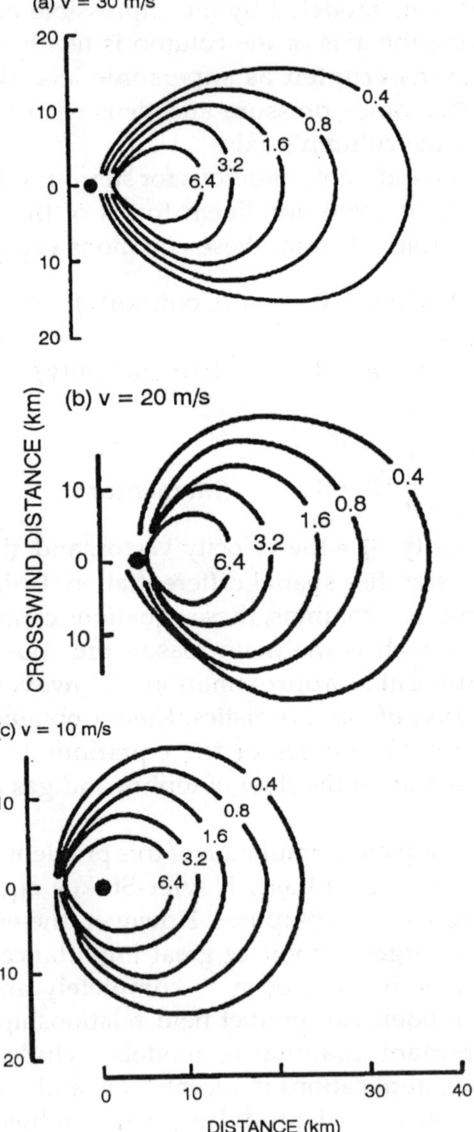

Fig. Plots of maximum clast isopleths

Systems associated with volcanoes. Progress towards these latter goals has been greatly aided by the development of a hydrovolcanism theory that links quantitative models of explosive eruption with the hydrological character of the volcano. Through this theory, both the heat resource and water necessary for a geothermal system can be simply assessed by characterization of explosive eruption products.

Hydrovolcanism

Hydrovolcanism is a broad term that encompasses the role of external (nonmagmatic) water in volcanic activity; synonyms include *phreatomagmatism* and *hydromagmatism*. This topic may have its roots in the 18th Century Neptunists' theory about the origin of basaltic rocks in oceans. After the eruption of Krakatau in 1883, world attention was focused on the dynamic potential of oceanic volcanism. Because water plays such a fundamental role in geothermal systems, we will briefly describe some research efforts that have unraveled the complexities of water/magma interactions in volcanic settings.

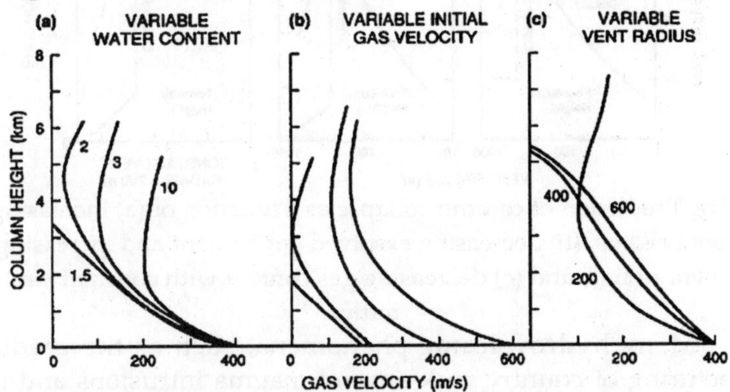

Fig. Velocity for the lower part of a Plinian eruption column.

This research has led to the development of systematics for inferring the existence of external water in volcanic areas. Such systematics concentrate on the interpretation of volcanic landforms and tephra deposits, which is viewed as a first step toward finding areas in which both a heat source and water exist. The study and characterization of hydrovolcanic features is chiefly used to make quick estimates of the abundance of water in a hydrothermal system. Detailed studies of water/magma interaction constrain subsurface conditions that have evolved within a geothermal system; for example, depth and lithology of aquifers and permeable formations, temperature of hydrothermal alteration, and spatial and temporal variations in subsurface hydrothermal behaviour.

A host of natural phenomena are produced by the interaction of magma or magmatic heat with an external source of water. Because the earth's crust is, in general, saturated with water, most volcanic fields have at least one feature produced by hydrovolcanic phenomena. Most widely recognized are phreatomagmatic and phreatic explosions.

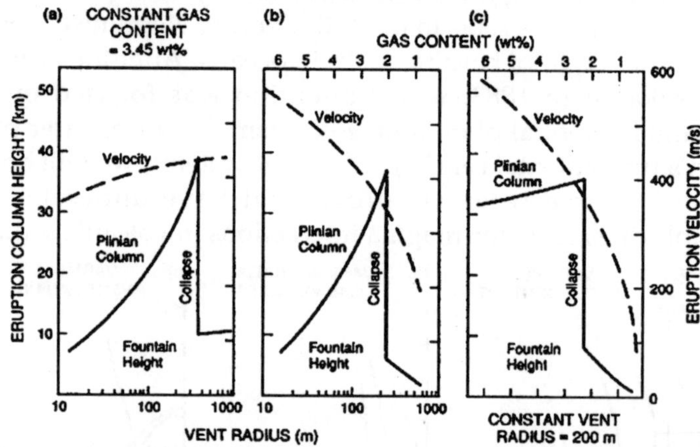

Fig. Prediction of column collapse as a function of (a) increasing vent radius, (b) decreasing exsolved gas content and increasing vent radius, and (c) decreasing gas content with constant vent radius.

Many hydrovolcanic phenomena, such as the gradual fracturing of country rock around magma intrusions and the alteration of rocks in hydrothermal systems are neither explosive nor readily observable. In their review of hydrovolcanism, various aspects of research, including:

- Geologic environments where systems occur,
- The range of physical phenomena,
- The wide variety of classical eruption types and landforms,

- Experimental modeling,
- Petrography of hydrovolcanic products,
- Textural analysis and indicators of water abundance in deposits, and
- Hydrovolcanic cycles.

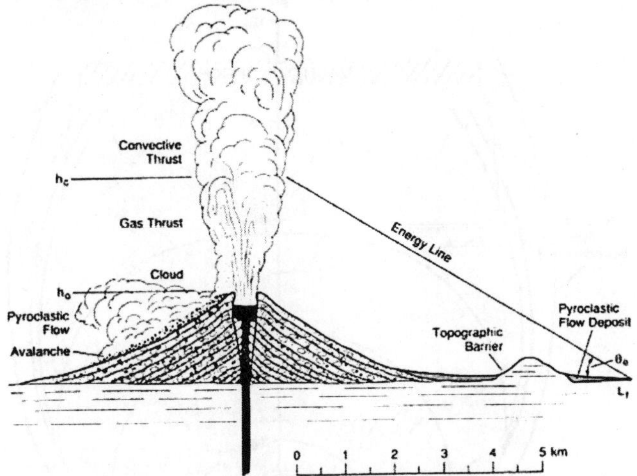

Fig.hypothetical composite cone.

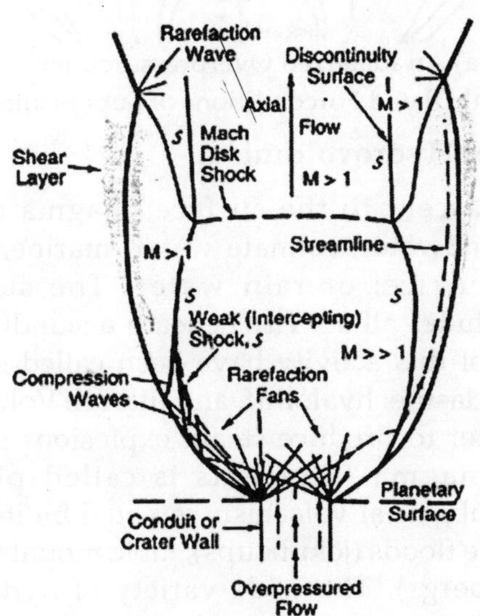

During recent years, hydrovolcanism has developed as a

field theory that applies to a range of physical as well as chemical processes (for example, magma differentiation by fluid and vapour transport, dynamic magma alteration during eruption, and contamination of magma bodies by external water).

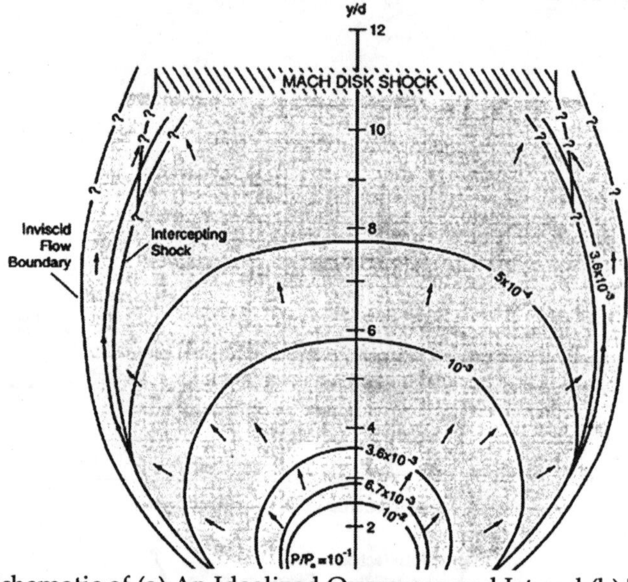

Fig. Schematic of (a) An Idealized Overpressured Jet and (b) The jet Structure Calculated Forconditions of Supersonic Flow

Environments of Hydrovolcanism

During its ascent to the surface, magma commonly encounters groundwater; connate water; marine, fluvial, or lacustrine water; ice; or rain water. The subaqueous environment includes all activity beneath a standing body of water products of this activity have been called subaquatic, aquagene, hyaloclastite, hyalotuff, and littoral. Volcanism that heats groundwater to produce steam explosions that do not eject juvenile magma fragments is called phreatic or hydrothermal. Subglacial volcanism is noted by its products, including massive floods (jökullaups), table mountains (stapi), and ridges (mobergs). The wide variety of hydrovolcanic phenomena underscores the fact that interaction between water and magma or magmatic heat should be expected in any volcanic setting. One long-held theory suggests that the depth below surface at which dynamic, water/magma interaction is possible is limited by the critical pressure of water or water-rich fluids, and that above this pressure, the phase change from liquid to gas upon heating does not involve large-volume changes. Accordingly, depths of 0.8 to 2.2 km were considered limits to explosive magma/water interaction. However, more recent work suggests that the critical point need not be a limitation to dynamic interaction and that expansion of water through its two-phase field is not required for rapid volume changes (Wohletz, 1986).

Nature of Hydrovolcanic Phenomena

The physical phenomena of hydrovolcanism belong to a class of well-studied physical processes termed fuel-coolant interactions (FCI). Depicts a hypothetical geologic system in which magma (fuel) explosively interacts with water-saturated sediments (coolant). This process occurs in stages of:

- Initial contact and steam-film development,
- Coarse mixing of magma and water or water-rich rock,
- Vapour expansion and flow, and finally
- Explosion and fine fragmentation of the magma. The process does not necessarily evolve through all these stages and may be arrested, for instance, before mixing or explosion.

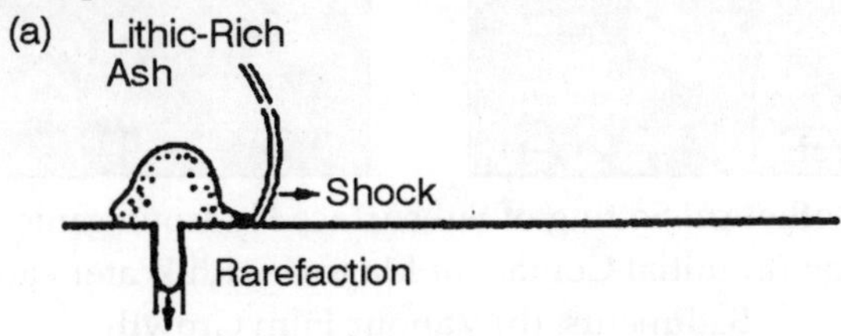

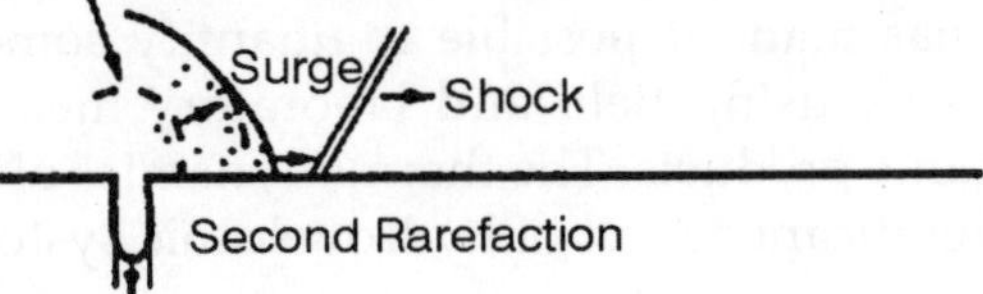

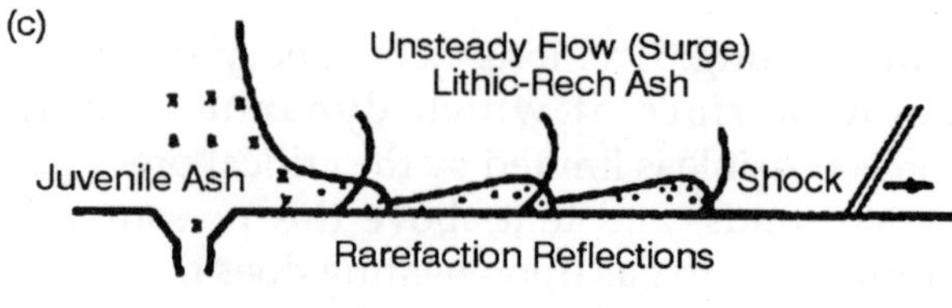

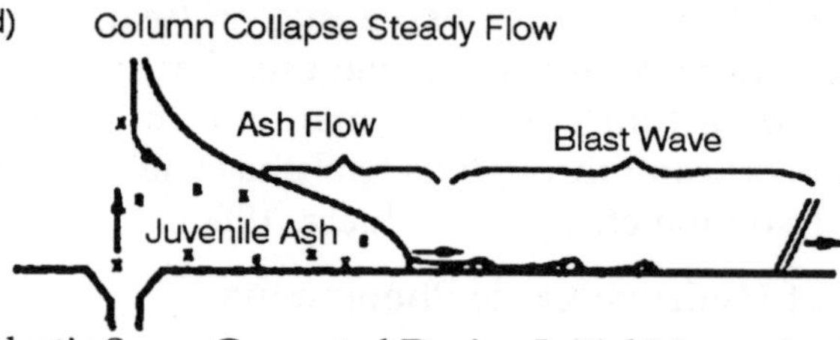

Fig. Pyroclastic Surge Generated During Initial Moments of Plinian Eruption. (a) A Rarefaction wave Recedes into the Magma Reservoir. (b and c) Reflected Rarefactions from the Reservoir and Flow Argins form Weak Shocks that Accelerate Ash in Surges. (d) The Flow of Juvenile Ash.

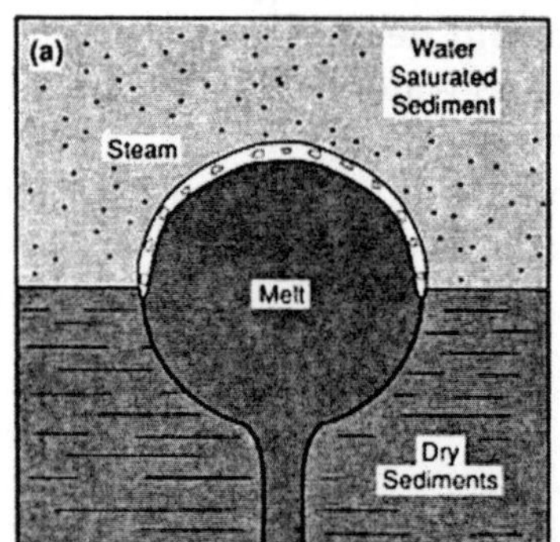

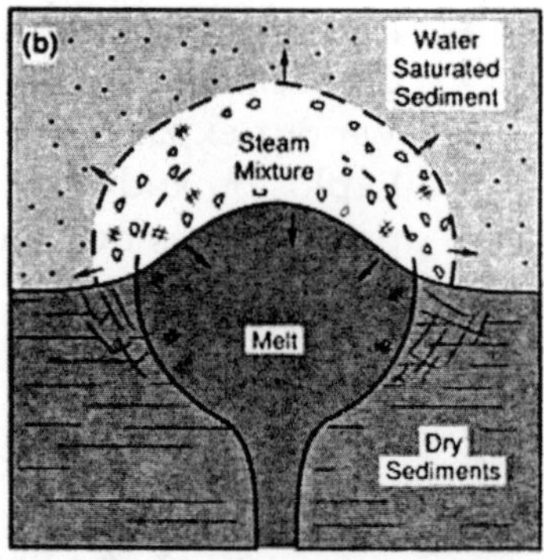

Fig. Hypothetical Setting of Subsurface Hydrovolcanic Activity, Showing (a) Initial Contact of Magma with Water-saturated Sediments, (b) Vapour Film Growth.

Much of our theoretical-understanding of hydrovolcanism has developed from laboratory experiments. This approach has made it possible to quantify some controlling parameters by using field and laboratory measurements of hydrovolcanic products. The thermodynamics of heat transfer is also a significant aspect of hydrovolcanic systems and their physical and chemical effects. The mechanical work produced by interaction of magma with external water is partitioned into

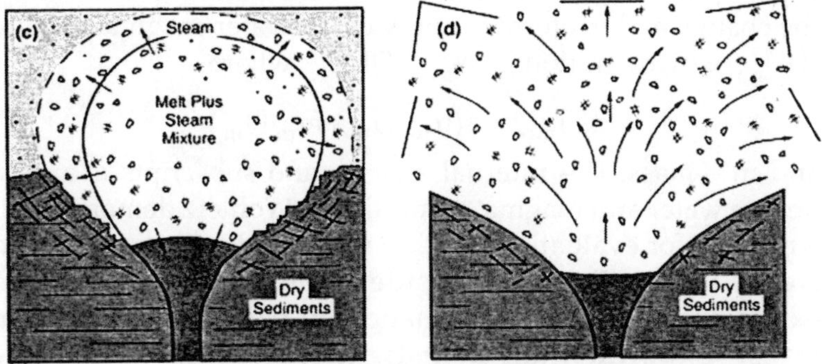

Fig. (b) vapour film growth, (c) mixing of magma with the sediments, and (d) expansion of the high-pressure steam in an explosion.

many possible modes, including fragmentation of the magma and country rock; excavation of a crater; dispersal of tephra; seismic and acoustic perturbations; and chemical processes such as solution and precipitation, mass diffusion, and magma quenching and crystallization.

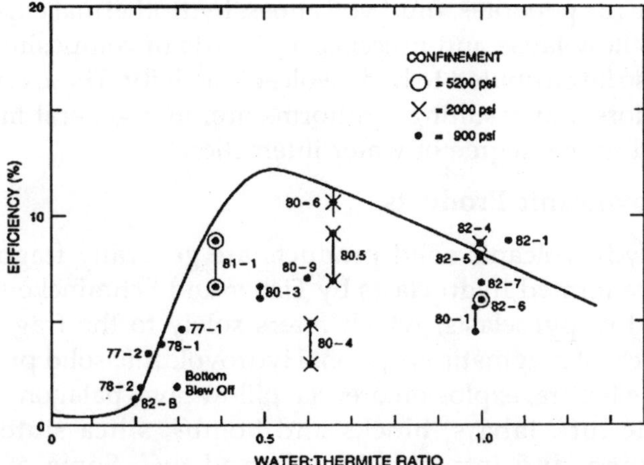

This work, $D\ W_{sys}$, is the sum of changes in kinetic energy ($D\ E_k$), potential energy (DE_p), and volume-change work ($pD\ V_{sys}$), which is given by

$$\Delta W_{sys} = \Delta KE + DPE + p\Delta V_{sys}.$$

There are several ways by which to evaluate the above expression, but one of the most direct methods is to estimate the change in the internal energy of the water/magma mixture (DU_{mix}); by definition $D\ W_{sys}{}^{\circ}$-DU_{mix}, where

$$\Delta U_{mix} = \Delta(m_w U_w + m_m U_m)$$

and m = mass, U = internal energy, and subscripts *w* and *m* denote water and magma, respectively. Wohletz demonstrated a method for evaluating Eq. that requires data from extended steam tables. Further consideration of the mixing and explosion stage yields information on particle velocities and sizes. Experimental investigations of water/magma interactions have displayed a variety of explosive and nonexplosive behaviors that are analogous to natural volcanic activity. These results support observations of hydrovolcanic eruptions in which a wide variety of classical eruption types have involved external water.

There is additional evidence of hydrovolcanism in a variety of landforms that range from small maar/tuffring craters to some large caldera outflow sheets of tephra. Such features as peperites, mud volcanoes, hydrothermal explosion pits, pillow lavas and breccias, and parts of composite cones can also be attributed to hydrovolcanic activity. These eruptive behaviors and resulting landforms are, in a general fashion, related to the degree of water interaction.

Hydrovolcanic Products

Hydrovolcanic solid products are generally fragmental and are termed hydroclasts by Fisher and Schmincke (1984), instead of pyroclasts, which refers solely to the fragmental products of magmatic eruption. Hydrovolcanic solid products include tephra, explosion breccia, pillow lava, palagonitic and zeolitic tuff, lahars, blocks and bombs, silica sinter and travertine, and intrusive breccia and tuff. Some of these materials involve posteruptive processes (for example, hydrothermal) in which water interacts with volcanic products. Petrographic studies of hydrovolcanic products involve determining the grain-size and textures of tephra and the chemical signatures caused by rapid and slow alteration. These data are indicators of the degree and type of water interaction. For example, the grain size of hydroclasts is a function of the mass ratio of interacting water and magma; grain textures are indicative of the type of interaction—passive, explosive, extensive, or transient. Field characterization of hydroclastic products focuses on:

- Analysis of various ejecta deposit characteristics, including textural analysis of bedforms, lithification, and deposit thickness *vs* distance from the vent, and
- Correlation of these observations with vent type.

A correlation can be made between the median grain diameters of hydrovolcanic products and the water/magma mass ratio this correlation was developed from both experimental and field applications. In general, hydrovolcanic tephra are distinguishable from magmatic tephra by their much finer grain size. Microscopic examination of grain shapes and textures also reveals hydrovolcanic features. Quantitative analyses of these features can document the relative importance of hydrovolcanic (wet) and magmatic (dry) mechanisms in samples from deposits of mixed origins. Hydrovolcanic grain textures are also indicative of the type of water/magma interaction.

Hydrovolcanic Cycles and Geothermal Energy

Hydrovolcanic phenomena occur in regular patterns at some volcanoes and thus can assist in defining cycles that in turn are useful in both predictions of future activity and estimates of subsurface hydrological conditions. The eruptive cycles portrayed in for example, show the changing availability of groundwater during periods of activity at several volcanoes. Cycles can be documented by careful field and laboratory analyses of volcanic products in which the abundance of erupted steam and its temperature are constrained by textural indicators of grain cohesion, deposit mobility as a function of moisture abundance, and degree of clast alteration.

Cycles are characterized as "wet" when the volcanic products indicate an increase of water during the eruptions;

"dry" cycles produce tephra that indicate decreasing water abundance throughout the eruption. The nature of these water indicators also demonstrates whether the erupted steam is saturated (wet) or super-heated (dry).

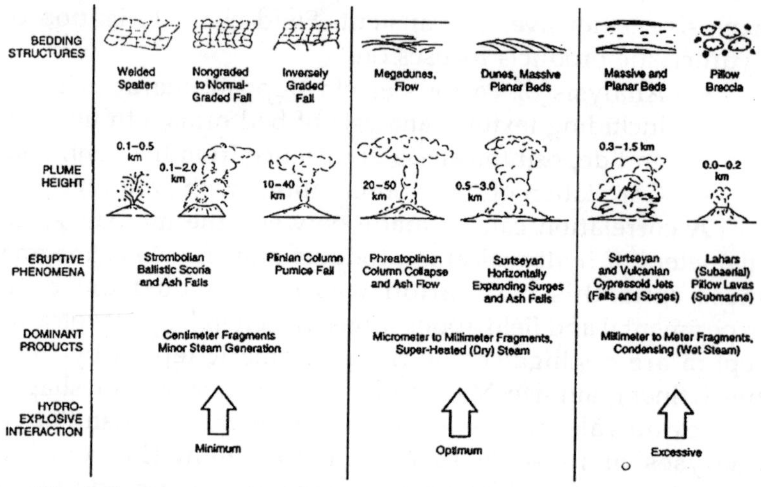

Fig. Relationship of Eruptive Phenomena, Deposit Type, and Landform to Water-to-magma Interaction Ratio.

As a general rule, locations that show wet cycles might be better candidates for geothermal exploration because they prove that water is sufficiently abundant in the volcanic system to quench the magma to water-vaporization temperatures.

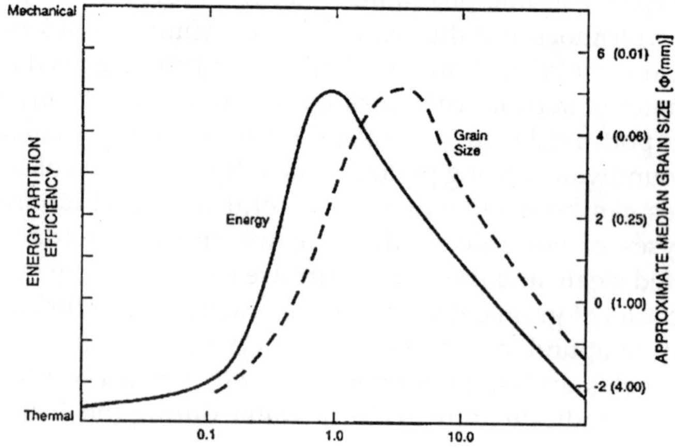

When estimating the volume of erupted hydroclastic products, this general rule constrains the volume of water involved in the eruptions and thus provides a measure of water abundance in the volcanic system.

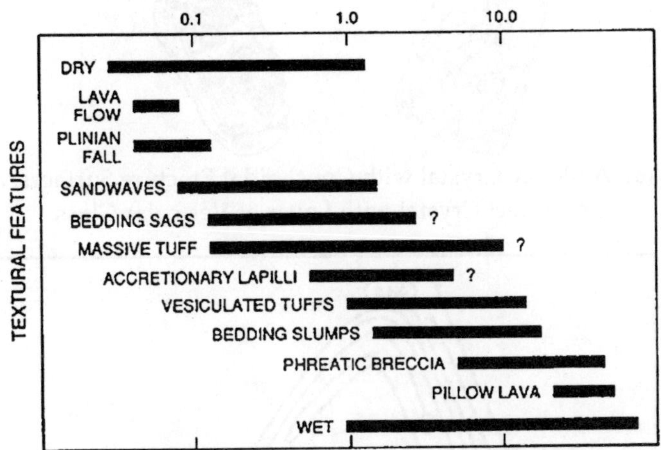

Fig. Correlation of Deposit Texture and Grain size to Water-to-Magma Ratio.

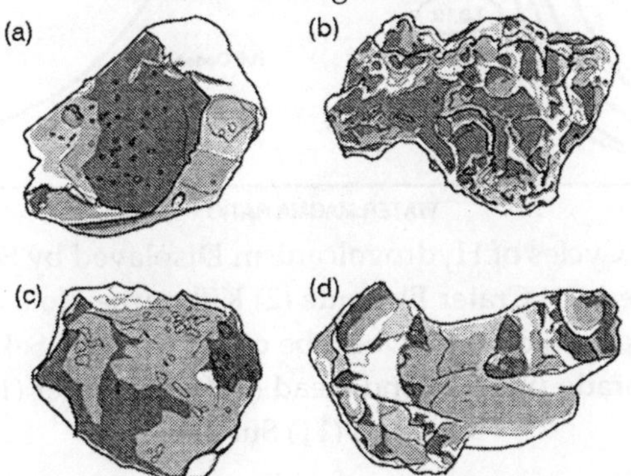

Fig. Sketches of Pyroclast Textures Resulting from Hydrovolcanism. (a) A Characteristic Blocky and Equant Glass Shard, (b) A Vesicular Grain Shard with Cleaved Vesicle Surfaces, (c) A Platy Shard, (d) A Drop-like or Fused Shard

The correlation between geothermal localities and phreatomagmatic volcanoes in Italy, especially those showing wet cycles. In addition, these authors demonstrated how the study of phreatomagmatic products helps locate and characterize a geothermal reservoir with respect to its lithology and fracture permeability.

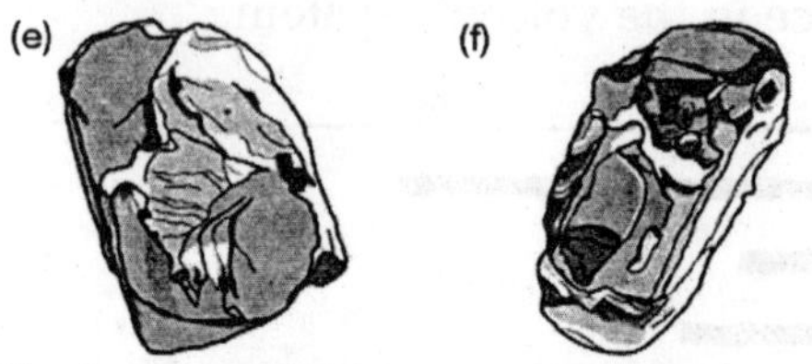

Fig. (e) A Blocky Crystal with Conchoidal Fracture Surfaces, and (f) A Perfect Crystal with Layer of Vesicular Glass.

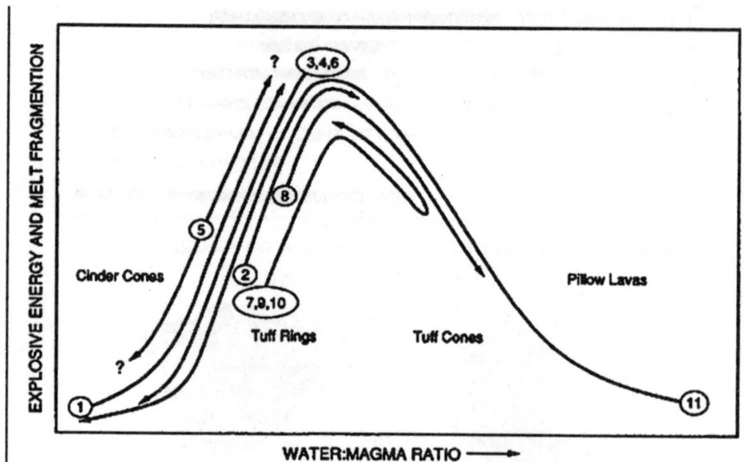

Fig. Various Cycles of Hydrovolcanism Displayed by Several Type of Volcanoes. (1) Crater Elegante (2) Kilbourne Hole (3) Peridot Mesa (4) Taal volcano (5) Ubehebe crater (6) Zuni Salt Lake, (7) Cerro Colorado (8) Diamond Head (9) Koko Crater (10) Pavant Butteand (11) Surtsey.

Chapter 7

Hydrothermal Energy

WATERPOWER

Waterpower, power derived from the fall of water from a higher to a lower level, and extracted by means of waterwheels or hydraulic turbines. Waterpower is a natural resource, available wherever a sufficient volume of steady water flow exists. The development of waterpower today requires extensive construction, including storage lakes, dams, bypass canals, and the installation of large turbines and electric generating equipment.

Because the development of hydroelectric power requires a large capital investment, it is often uneconomical for a region where coal or oil is cheap, even though the cost of fuel for a steam-powered generating plant is higher than the cost of running a hydroelectric plant. However, increasing environmental concerns are focusing attention on renewable energy sources such as water.

The use of waterpower dates from ancient Greece and Rome, where waterwheels were used for the milling of corn. The availability of cheap slave and animal labor, however, restricted the widespread application of waterpower until about the 12th century. During the Middle Ages, large wooden waterwheels were developed with a maximum power output of about 50 horsepower. Modern waterpower owes its development to the British civil engineer John Smeaton, who first built large waterwheels using cast-iron construction.

Early American waterpower works were limited to drops of about 5 m (about 16 ft), and dams and canals were necessary for the installation of successive waterwheels when the drop was greater. Large storage-dam construction, however, was not feasible, and low water flows during summer and fall, coupled with icing during the winter, led to the replacement of nearly all waterwheels by steam when coal became readily available.

HYDROPOWER

Hydropower is energy obtained from flowing water. Energy in water can be harnessed and used for this usefulness, in the form of motive energy or temperature differences. The most common application is the dam, but it can be used directly as a mechanical force or a thermal source/sink.

Prior to the widespread availability of commercial electricity, hydropower was widely used for milling, textile manufacture, and the operation of sawmills. In the 1830s, at the height of the canal-building era, hydropower was used to transport barge traffic up and down steep hills using the technology of inclined plane railroads.

Types of water power

There are many forms of water power:

- Waterwheels, used for hundreds of years to power mills and machinary
- Hydroelectric energy, a term usually reserved for hydroelectric dams.
- Tidal power, which captures energy from the tides in horizontal direction
- Tidal stream power, which does the same vertically
- Wave power, which uses the energy in waves

Hydroelectric power

Hydroelectric power from potential energy of the elevation of waters, now supplies about 715,000 MWe or 19% of world electricity, and large dams are still being designed. Apart from a few countries with an abundance of it, hydro power is normally applied to peak-load demand, because it is so readily stopped and started. Nevertheless, hydroelectric power is probably not a major option for the future of energy production in the developed nations because most major sites within these nations with the potential for harnessing gravity in this way are either already being exploited or are unavailable for other reasons such as environmental considerations. Hydroelectric energy produces essentially no carbon dioxide, in contrast to burning fossil fuels or gas, and so is not a significant contributor to global warming through CO_2. Recent reports have linked hydroelectric power to methane, which forms out of decaying submerged plants which grow in the dried up parts of the basis in times of drought. Methane is a greenhouse gas.

Hydroelectric power can be far less expensive than electricity generated from fossil fuel or nuclear energy. Areas with abundant hydroelectric power attract industry with low cost electricity. Recently, increased environmental concerns surrounding hydroelectric power, have begun to outweigh cheap electricity in some countries.

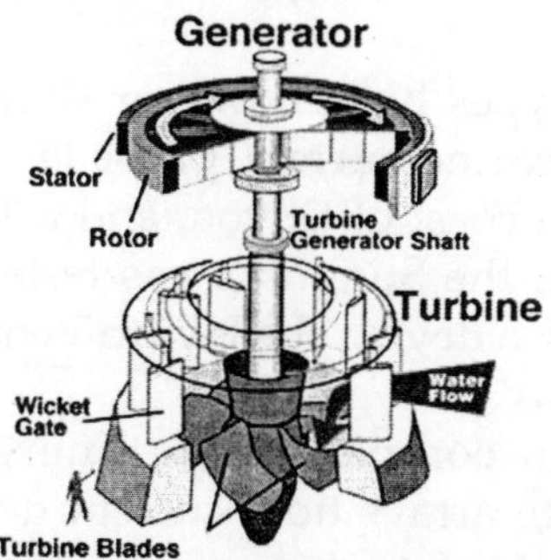

Fig. Hydraulic Turbine and Electrical Generator.

The chief advantage of hydroelectric dams is their ability to handle seasonal high peak loads. When the electricity demands drop, the dam simply stores more water. Some electricity generators use water dams to store excess energy by using the electricity to pump water up into a basin. The electricity can be re-generated when demand increases. In practice the utilization of stored water in river dams is sometimes complicated by demands for irrigation which may occur out of phase with peak electrical demands.

Tidal Power

The trapped water can be used to turn turbines as it is released through the tidal barrage in either direction. Worldwide this technology appears to have little potential, largely due to environmental constraints.Another possible fault is that the system would generate electricity most efficiently if it were to generate electricity in bursts, every six hours (once every tide). Obviously, this limits the applications for which tidal energy can be used.

Tidal Stream Power

A relatively new technology development, tidal stream generators draw energy from underwater currents in much the same way that wind generators are powered by the wind. The much higher density of water means that there is the potential for a single generator to provide significant levels of power. Tidal stream technology is at the very early stages of development though and will require significantly more research before it becomes a significant contributor to electrical generation needs.

Several prototypes have however shown some promise 300 kW Seaflow marine current propeller type turbine was tested off the north coast of Devon, and a 150 kW oscillating hydroplane device, the Stingray, was tested off the Scottish coast. Another British device, the Hydro Venturi, is to be tested in San Francisco Bay.

The Canadian company Blue Energy has plans for installing very large arrays tidal current devices mounted in what they call a 'tidal fence' in various locations around the world, based on a vertical axis turbine design.

Wave Power

Harnessing power from ocean surface wave motion is a possibility which might yield much more energy than tides. The feasibility of this has been investigated, particularly in the UK. Generators either coupled to floating devices or turned by air displaced by waves in a hollow concrete structure would produce electricity for delivery to shore. Numerous practical problems have frustrated progress. A prototype shore based wave power generator is being constructed at Port Kembla in Australia and is expected to generate up to 500 MWh per annum. The Wave Energy Converter has been constructed and initial test results have exceeded expectations in terms of energy production during times of low wave energy. The energy of waves crashing against the shore is absorbed by an air driven generator and converted to electricity. For countries with large coastlines and rough sea conditions the energy density of breaking waves offers the possibility of generating electricity in utility volumes. Excess power in periods of rough sea could be used to generate renewable hydrogen.

DEVELOPMENT OF HYDROELECTRIC POWER

The rebirth of waterpower had to await the development of the electric generator, improvement of the hydraulic turbine, and the growing demand for electricity by the turn of the 20th century. Commercial power companies began to install a large number of small hydroelectric plants in the mountain regions near the major population centers, and by 1920 hydroelectric plants accounted for 40 percent of the electric power produced in the United States.

Hydroelectric power generation was accelerated by the establishment of the Federal Power Commission in 1920. Although additional hydroelectric plants were being built, the simultaneous development of larger and more cost-efficient steam-power plants made it obvious that only very large and costly hydroelectric installations could compete effectively, and that the federal government would have to assume a major share in their construction.

Motivated by the search for the multiple use of water resources, including navigation, flood control, and irrigation, in addition to power production, the Tennessee Valley Authority, or TVA, started government participation in large-scale waterpower development in 1933.

Most major installations depend on a large water-storage reservoir upstream of the dam where water flow can be controlled and a nearly constant water level can be assured.

Water flows through conduits, called penstocks, which are controlled by valves or turbine gates to adjust the flow rate in line with the power demand.

The water then enters the turbines and leaves them through the so-called tailrace. The power generators are mounted directly above the turbines on vertical shafts. The design of turbines depends on the available head of water, with so-called Francis-type turbines used for high heads and Kaplan, or propeller turbines, used for low heads.

In contrast to storage-type plants, which depend on the impounding of large amounts of water, a few examples exist where both the water drop and the steady flow rate are high enough to permit so-called run-of-the-river installations; one such is the joint American-Canadian Niagara Falls power project.

Small-scale hydroelectric plants, with capacities between 1 kilowatt and 1 megawatt, are being developed in some countries. In many of China's districts, such dams are the main source of electric power. Other developing nations are also showing interest in such projects, which can make good use of available labor. In the United States, interest in small-scale plants increased following passage of the Public Utilities Regulatory Policies Act in 1978, because it states that large utilities must buy power fed into their lines by these small-scale plants.

GLOBAL HYDROELECTRIC POWER GENERATION

Worldwide, hydropower represented 17 percent of the total energy generated in 2003, the most recent year for which data are available. In many countries, hydroelectric power is the dominant source of electric power. In 2003 Norway derived 99 percent of its power from hydroelectric plants. The same year, hydroelectric power provided 100 percent of the electricity used in the Democratic Republic of the Congo (DRC, formerly Zaire) and 84 percent of the electricity used in Brazil.

The Itaipu hydroelectric plant on the Paraná River between Brazil and Paraguay, officially dedicated in 1982, has the greatest capacity in the world (12,600 megawatts when placed in full operation). The Grand Coulee Dam, the largest hydroelectric power plant in North America, provides about 6,480 megawatts.

Canada, the largest producer of hydroelectric power in the world, generated 332.5 billion kilowatt-hours (kwh) in 2003. This constituted 59 percent of the nation's electric power. Hydroelectric-power generation in the United States increased from about 16 billion kwh in 1920 to 275.8 billion kwh in 2003. Although the United States runs a close second to Canada in the total amount of hydroelectric power produced, only 7 percent of the electric power used in the United States was generated by hydroelectric power plants in 2003.

ENVIRONMENTAL IMPACT

Hydroelectric power is regarded by some as a relatively clean source of energy because it emits fewer greenhouse gases than thermal power plants. Greenhouse gases contribute to global warming and climate change, and for many environmentalists, the buildup of greenhouse gases in the atmosphere is the most important environmental issue. However, the dams used to generate hydroelectricity can have other adverse environmental consequences.

Dams alter the water temperatures and microhabitats downstream. Water released from behind dams usually comes from close to the bottom of the reservoirs, where little sunlight penetrates. This frigid water significantly lowers the temperatures of sun-warmed shallows downstream, rendering them unfit for certain kinds of fish and other wildlife. Natural rivers surge and meander, creating small pools and sandbars that provide a place for young fish, insects, and other river-dwelling organisms to flourish. But dams alter the river flow, eliminating these microhabitats and, in some cases, their inhabitants.

Finally, dams also prevent nutrient-laden silt from flowing downstream and into river valleys. Water in a fast-moving river carries tiny particles of soil and organic material.

As it slows, the organic matter it carries drops to the river bottom or accumulates along the banks. Following heavy rains or snowmelt, rivers spill over their banks and deposit organic matter on their floodplains, creating rich, fertile soil. Some of the organic matter makes it all the way to river mouths, where it settles into the rich mud of estuaries, ecosystems that nourish up to one-half of the living matter in the world's oceans. Large dams artificially slow water to a near standstill, causing the organic matter to settle to the bottom of the reservoir. In such cases, downstream regions are deprived of nutrient-laden silt.

WATERWHEEL

Waterwheel, simple mechanical device used to convert water power into rotary motion. A waterwheel works on the principle that when water undergoes a controlled change in elevation, the falling water is a source of power and will turn the axle of the waterwheel. The three general types of waterwheels, which are sometimes called gravity wheels, are the undershot, the overshot, and the breast wheel.

The oldest variety is the undershot wheel. This waterwheel was used by the early Egyptians and Persians to drive water-lifting devices, called norias, which were used in irrigating fields. The simplest form of an undershot waterwheel is a paddle wheel, which is placed with the lower paddles in a stream of water. The water current exerts a force against the paddles and turns the wheel. In order to improve the power obtained from a given source, the paddles may be operated in a trough that prevents water from passing around the paddles without doing work. The waterwheel was subsequently adapted to turn millstones for grinding corn and grain.

When a waterwheel is constructed in a hilly region where there is an appreciable difference between upstream and downstream water elevations, an undershot wheel with a regulating gate or a breast waterwheel may be used. A breast waterwheel is located at the base of a waterfall; the water catches and turns the paddles as the water flows over the falls. Various forms of the undershot wheel were used exclusively until the time of the Renaissance. The early water wheels were constructed of wood; use of iron parts became common during the Renaissance. In areas where the difference in water elevation is about equal to the wheel diameter or slightly larger, the more efficient overshot wheel is installed. In the overshot wheel, which came into common use during the Renaissance, the stream of water is directed to flow over the waterwheel paddles, and additional energy is gained from the falling water.

With the introduction of hydraulic turbines in the early part of the 19th century, the importance of waterwheels decreased. Today they are limited to small power plants. The best modern overshot waterwheels attain efficiencies of 85 percent or greater, breast wheels about 75 percent, and undershot wheels about 35 percent.

TURBINE

Turbine, rotary engine that converts the energy of a moving stream of water, steam, or gas into mechanical energy. The basic element in a turbine is a wheel or rotor with paddles, propellers, blades, or buckets arranged on its circumference in such a fashion that the moving fluid exerts a tangential force that turns the wheel and imparts energy to it. This mechanical energy is then transferred through a drive shaft to operate a machine, compressor, electric generator, or propeller. Turbines are classified as hydraulic, or water, turbines, steam turbines, or gas turbines. Today turbine-powered generators produce most of the world's electrical energy. Windmills that generate electricity are known as wind turbines.

Hydraulic Turbines

The oldest and simplest form of the hydraulic turbine was the waterwheel, first used in ancient Greece and subsequently adopted in most of ancient and medieval Europe for grinding grain. It consisted of a vertical shaft with a set of radial vanes or paddles positioned in a swiftly flowing stream or millrace. Its power output was about 0.5 horsepower. The horizontal-shaft waterwheel (that is, a horizontal shaft connected to a vertical paddle wheel), first described by the Roman architect and engineer Marcus Vitruvius Pollio during the 1st century

BC, had the lower segment of the paddle wheel inserted into the stream, thus acting as a so-called undershot waterwheel. By about the 2nd century AD, the more efficient overshot wheel had come into use in hilly regions.

Here the water was poured on the paddles from, and additional energy was gained from the falling water. The maximum power of the waterwheel, which was constructed of wood, increased from about 3 horsepower to about 50 horsepower in the Middle Ages.

The transition from waterwheel to turbine is largely semantic. The first important attempt to formulate a theoretical basis for waterwheel design was in the 18th century by the British civil engineer John Smeaton, who proved that the overshot wheel was more efficient.

The French military engineer Jean Victor Poncelet, however, devised an undershot wheel, the curved blades of which raised efficiency to nearly 70 percent; it quickly came into wide use.

Another French military engineer, Claude Burdin, invented the term *turbine,* introduced as part of which he stressed speed of rotation. Benoit Fourneyron, who studied under Burdin at the School of Mines at St. Étienne, designed and built wheels that achieved speeds of 60 or more rpm (revolutions per minute) and provided up to 50 horsepower for French ironworks. Ultimately Fourneyron built turbines that operated at 2300 rpm, developing 60 horsepower at an efficiency of more than 80 percent.

Despite its remarkable efficiency, the Fourneyron turbine had certain drawbacks as a result of the radial outward flow of water that passed through it. This created problems if water flow was reduced or load removed. The British-born American engineer James B. Francis designed a turbine in which the flow was inward, and the so-called reaction, or Francis, turbine, became the most widely used hydraulic turbine for water pressures, or heads, equivalent to a column of water 10 to 100 m (33 to 330 ft). This type of turbine operates by expanding the pressure energy in the water during the flow through the blade passages, resulting in a net force, or reaction, which has a tangential component that turns the wheel. For installations where water heads of about 90 to 900 m (about 300 to 3000 ft) were available, the Pelton wheel, named after the American engineer Lester Allen Pelton, came into use during the second half of the 19th century.

In this turbine, the water is piped from a high-level reservoir through a long duct, or penstock, to a nozzle where its energy is converted into the kinetic energy of a high-speed jet. This jet is then directed onto curved buckets, which turn the flow by nearly 180 degrees and extract the momentum. Because the action of the Pelton wheel depends on the impulse of the jet on the wheel, rather than on the reaction of the expanding water, this type of turbine is also known as an impulse turbine.

The increasing demand for hydroelectric power during the early 20th century led to the need for a turbine suitable for small water heads of 3 to 9 m (about 10 to 30 ft) that could be employed in many rivers where low dams could be built. In 1913 the Austrian engineer Viktor Kaplan first proposed his propeller (or Kaplan) turbine, which basically acts like a ship's propeller in reverse. Kaplan later improved his turbine by allowing the blades to swivel about their axis. These variable pitch propellers improved efficiency by optimally matching the blade angle to the head, or flow rate.

To maintain constant output voltage in a hydroelectric installation, the turbine speed must be kept constant regardless of variations in the water pressure acting on it. This requires extensive controls that, for both Francis and Kaplan turbines, act primarily to open or close the guide-vane passages to regulate the flow and, in the case of Kaplan turbines, to vary the pitch of the propellers.

In a Pelton wheel installation, the water flow is adjusted by opening or closing the supply nozzles. Here a temporary spill bypass nozzle has to be provided since rapid flow changes in long penstocks would induce pressure surges, called water hammers, which can be highly destructive. During adjustments, the total water flow through both the supply and spill nozzles must be kept nearly constant with the eventual closing of the bypass nozzle, which must be carried out very slowly to avoid water hammer.

Advances in Turbine Design

The trend in modern hydraulic turbine installations has been toward higher heads and larger units. Depending on the size of the unit, Kaplan turbines are now used with heads up to about 60 m (about 200 ft), and Francis turbines up to 610 m (2000 ft). The world's highest head installation (about 1770 m/ about 5800 ft), using a Pelton wheel, is at Reisseck, Austria, and the largest single units are installed in a plant at Itaipu, Brazil, where 18 Francis-type turbines sized at 700 megawatts (Mw) each have a total capacity of 12,600 Mw. The largest installation in North America is at La Grande on James Bay, in eastern Canada, where 22 units rated 333 Mw each have a total capacity of 7326 Mw.

Many of the small dam hydroelectric systems built before 1930 were later abandoned because of high maintenance and labor costs. Increases in the cost of fossil fuels have focused renewed attention on these low head installations. With the development of standardized propeller turbines with nearly horizontal shafts, small installations have again become attractive.

Turbines can also be designed to run in reverse as pumps. This is done by inverting the electric generator to operate as a motor. Because electric power cannot be stored economically, the operation of the so-called pump-turbines with electricity generated from nuclear and fossil fuel power plants during off-peak hours enables additional water to be stored in a reservoir. It can then be reused to drive the turbine during peak periods. In recent years, pump-turbine technology has been developed to allow for heads up to about 600 m (about 2000 ft) of water and for unit capacities of more than 400 Mw.

STEAM TURBINES

The success of the water turbine inevitably led to consideration of the turbine principle for extracting power from steam. Where the Watt-type reciprocating steam engine utilized the pressure of steam, the turbine could achieve higher efficiency by utilizing the kinetic energy of steam flow. The turbine can be made smaller, lighter, and cheaper than a reciprocating steam engine of comparable power. It can be made in far larger sizes than the conventional steam engine. Mechanically, it has the advantage of producing rotating motion directly without the necessity of using a crankshaft or other means of transforming reciprocal to rotary motion. As a result, the steam turbine has supplanted the reciprocating engine as a prime mover in large electricity-generating plants and is also used as a means of jet propulsion.

Steam turbines are used in the generation of nuclear power and in nuclear ship propulsion. They operate with fuel-fired boilers for power generation. In cogeneration applications requiring both process heat (heat used in an industrial process) and electricity, steam is raised at high pressure in the boiler and extracted from the turbine at the pressure and temperature required by the process.

Steam turbines may be used in combined cycles with a steam generator which recovers heat that would otherwise be lost. Industrial units are used to drive machines, pumps, compressors, and electrical generators. Ratings range from a few horsepower to more than 1300 Mw.

The steam turbine was not invented by any one individual but was the result of work by a number of inventors in the latter part of the 19th century. Notable contributors to the development of the turbine were the British inventor Charles Algernon Parsons and the Swedish inventor Carl Gustaf Patrik de Laval. Parsons was responsible for the so-called principle of staging, whereby steam was permitted to expand in a number of stages, performing useful work at each stage. De Laval was the first to design suitable jets and blades for the efficient use of the expanding steam.

The action of the steam turbine is based on the thermodynamic principle that when a vapour is allowed to expand, its temperature drops, and its internal energy is thereby decreased. This reduction in internal energy is transformed into mechanical energy in the form of an acceleration of the particles of the vapour. This transformation makes a large amount of work energy directly available. In the case of expanding steam, a reduction of 100 Btu in internal energy through expansion can result in increasing the speed of the steam particles to a rate of almost 2900 km/h (almost 1800 mph). At such speeds the energy available is great, even though the particles are extremely light.

Although they are built according to two different principles, the essential parts of all steam turbines are similar. They consist of nozzles or jets through which steam flows and expands, dropping in temperature, and gaining kinetic energy, and blades against which the swiftly moving steam exerts pressure. The arrangement of jets and blades, whether fixed or stationary, depends upon the type of turbine. In addition to these two basic components, turbines are equipped with wheels or drums upon which the blades are mounted, a shaft for these wheels or drums, an outer casing that confines the steam to the area of the turbine proper, and various pieces of auxiliary equipment, including lubrication devices and governors.

The simplest form of steam turbine is the so-called impulse turbine, in which the turbine jets are fixed in place on the inside of the turbine casing, and the blades are set on the rims of revolving wheels mounted on a central shaft. Steam passing through a fixed nozzle passes over the curved blades; these absorb some of the kinetic energy of the expanded steam, turning the wheel and shaft on which they are mounted. The turbine is designed so that steam entering at one end of the turbine expands through a succession of nozzles until it has lost most of its internal energy.

In the reaction turbine, mechanical energy is obtained to some degree by the impact of steam upon the blades, but primarily it is obtained by the acceleration of the steam as it expands. A turbine of this type consists of a set of fixed and a set of movable blades. The blades are arranged so that each pair acts as a nozzle through which the steam expands as its passes. The blades of a reaction turbine are usually mounted on a drum and not on a wheel. This drum acts as the shaft of the turbine. In order to use the energy available in steam efficiently in a turbine of either type, it is necessary to employ a number of stages, in each of which a small amount of thermal energy is converted to kinetic energy. If the entire conversion of energy took place instead in a single expansion stage, the rotative speed of the turbine wheel would be excessive.

In general, reaction turbines require more stages than impulse turbines. It can be shown that for the same diameter and energy range, a reaction turbine requires twice the number of stages for peak stage efficiency. Large turbines that are nominally of the impulse variety employ some reaction at the root of the steam path to assure efficient flow through the buckets. Many turbines that are nominally reactive have an impulse control stage first, which allows for a saving in the total number of stages.

Because of the increase in volume as the steam expands through the various stages of a turbine, the size of the openings through which the steam passes must increase from stage to stage. In the practical engineering design of turbines, this increase is accomplished by lengthening the blades from stage to stage and by increasing the diameter of the drum or wheel upon which the blades are mounted and by adding two or more turbine sections in parallel.

As a result, a small industrial turbine may be more or less conical in shape, with its smallest diameter at the high-pressure, or inlet, end, and its largest at the low-pressure, or exhaust, end. A large unit for a nuclear power station may have four rotors consisting of one double-flow high-pressure section followed by three double-flow low-pressure sections.

Impulse turbines usually employ pressure or Rateau staging, named after the French engineer Auguste Rateau, in which the pressure ratio across each stage is nearly uniform. Impulse turbines built in the past have made use of velocity-compounded, or Curtis, staging named after its American inventor, Charles Gordon Curtis, which has two sets of moving buckets with an intermediate set of fixed blades following the nozzles. The staging of a reaction turbine may be called Parsons' staging, after its British inventor, Charles Parsons.

Steam turbines are comparatively simple machines, having only one major moving part, the rotor; however, auxiliary equipment is necessary for their operation. Journal bearings support the shaft. A thrust bearing positions the shaft axially. An oil system provides lubrication to the bearings. Seals minimize steam leakage within the steam path. A sealing system prevents steam leaking from the machine and air leaking from the outside into the machine.

The speed of rotation is controlled by valves at the inlet(s) of the machine. In addition, reaction turbines develop considerable axial thrust owing to the pressure drop across the moving blades. This is usually compensated for by the use of a dummy piston, which creates a thrust in the opposite direction to that of the steam path.

The expansion efficiency of a modern multistage steam turbine is inherently high because of the state of development of the steam-path components and the ability to recover losses of one stage in those downstream through reheating. The efficiency with which a section of a turbine converts the theoretically available thermodynamic energy to mechanical work is commonly in excess of 90 percent. The thermodynamic efficiency of a steam-power installation is much less, owing to the energy lost in the exhaust steam from the turbine.

ELECTRIC POWER SYSTEMS

Electric Power Systems, components that transform other types of energy into electrical energy and transmit this energy to a consumer. The production and transmission of electricity is relatively efficient and inexpensive, although unlike other forms of energy, electricity is not easily stored and thus must generally be used as it is being produced.

Components of an Electric Power System

A modern electric power system consists of six main components:

- The power station,
- A set of transformers to raise the generated power to the high voltages used on the transmission lines,

- The transmission lines,
- The substations at which the power is stepped down to the voltage on the distribution lines,
- The distribution lines, and
- the transformers that lower the distribution voltage to the level used by the consumer's equipment.

Power Station

The power station of a power system consists of a prime mover, such as a turbine driven by water, steam, or combustion gases that operate a system of electric motors and generators. Most of the world's electric power is generated in steam plants driven by coal, oil, nuclear energy, or gas. A smaller percentage of the world's electric power is generated by hydroelectric (waterpower), diesel, and internal-combustion plants.

Transformers

Modern electric power systems use transformers to convert electricity into different voltages. With transformers, each stage of the system can be operated at an appropriate voltage. In a typical system, the generators at the power station deliver a voltage of from 1,000 to 26,000 volts (V). Transformers step this voltage up to values ranging from 138,000 to 765,000 V for the long-distance primary transmission line because higher voltages can be transmitted more efficiently over long distances. At the substation the voltage may be transformed down to levels of 69,000 to 138,000 V for transfer on the distribution system. Another set of transformers step the voltage down again to a distribution level such as 2,400 or 4,160 V or 15, 27, or 33 kilovolts (kV). Finally the voltage is transformed once again at the distribution transformer near the point of use to 240 or 120 V.

Transmission Lines

The lines of high-voltage transmission systems are usually composed of wires of copper, aluminum, or copper-clad or aluminum-clad steel, which are suspended from tall latticework towers of steel by strings of porcelain insulators. By the use of clad steel wires and high towers, the distance between towers can be increased, and the cost of the transmission line thus reduced. In modern installations with essentially straight paths, high-voltage lines may be built with as few as six towers to the kilometer. In some areas high-voltage lines are suspended from tall wooden poles spaced more closely together.

For lower voltage distribution lines, wooden poles are generally used rather than steel towers. In cities and other areas where open lines create a safety hazard or are considered unattractive, insulated underground cables are used for distribution. Some of these cables have a hollow core through which oil circulates under low pressure. The oil provides temporary protection from water damage to the enclosed wires should the cable develop a leak. Pipe-type cables in which three cables are enclosed in a pipe filled with oil under high pressure (14 kg per sq cm/200 psi) are frequently used. These cables are used for transmission of current at voltages as high as 345,000 V (or 345 kV).

Supplementary Equipment

Any electric-distribution system involves a large amount of supplementary equipment to protect the generators, transformers, and the transmission lines themselves. The system often includes devices designed to regulate the voltage or other characteristics of power delivered to consumers.

To protect all elements of a power system from short circuits and overloads, and for normal switching operations, circuit breakers are employed.

These breakers are large switches that are activated automatically in the event of a short circuit or other condition that produces a sudden rise of current. Because a current forms across the terminals of the circuit breaker at the moment when the current is interrupted, some large breakers (such as those used to protect a generator or a section of primary transmission line) are immersed in a liquid that is a poor conductor of electricity, such as oil, to quench the current. In large air-type circuit breakers, as well as in oil breakers, magnetic fields are used to break up the current. Small air-circuit breakers are used for protection in shops, factories, and in modern home installations. In residential electric wiring, fuses were once commonly employed for the same purpose. A fuse consists of a piece of alloy with a low melting point, inserted in the circuit, which melts, breaking the circuit if the current rises a certain value. Most residences now use air-circuit breakers.

POWER FAILURES

In most parts of the world, local or national electric utilities have joined in grid systems. The linking grids allow electricity generated in one area to be shared with others. Each utility that agrees to share gains an increased reserve capacity, use of larger, more efficient generators, and the ability to respond to local power failures by obtaining energy from a linking grid.

These interconnected grids are large, complex systems that contain elements operated by different groups. These systems offer the opportunity for economic savings and improve overall reliability but can create a risk of widespread failure.A major grid-system breakdown occurred on November 9, 1965, in eastern North America, when an automatic control device that regulates and directs current flow failed in Queenston, Ontario, causing a circuit breaker to remain open. A surge of excess current was transmitted through the northeastern United States.

Generator safety switches from Rochester, New York, to Boston, Massachusetts, were automatically tripped, cutting generators out of the system to protect them from damage. Power generated by more southerly plants rushed to fill the vacuum and overloaded these plants, which automatically shut themselves off. The power failure enveloped an area of more than 200,000 sq km (80,000 sq mi), including the cities of Boston; Buffalo, New York; Rochester, New York; and New York City.

Similar grid failures, usually on a smaller scale, have troubled systems in North America and elsewhere. On July

13, 1977, about 9 million people in the New York City area were once again without power when major transmission lines failed. In some areas the outage lasted 25 hours as restored high voltage burned out equipment. These major failures are termed blackouts.

The worst blackout in the history of the United States and Canada occurred August 14, 2003, when 61,800 megawatts of electrical power was lost in an area covering 50 million people. (One megawatt of electricity is roughly the amount needed to power 750 residential homes.) The blackout affected such major cities as Cleveland, Detroit, New York, Ottawa, and Toronto. Parts of eight states—Connecticut, Massachusetts, Michigan, New Jersey, New York, Ohio, Pennsylvania, and Vermont—and the Canadian provinces of Ontario and Québec were affected. The blackout prompted calls to replace aging equipment and raised questions about the reliability of the national power grid.

The term *brownout* is often used for partial shutdowns of power, usually deliberate, either to save electricity or as a wartime security measure. From November 2000 through May 2001 California experienced a series of planned brownouts to groups of customers, for a limited duration, in order to reduce total system load and avoid a blackout due to alleged electrical shortages. However, an investigation by the California Public Utilities Commission into the alleged shortages later revealed that five energy companies withheld electricity they could have produced. In 2002 the commission concluded that the withholding of electricity contributed to an "unconscionable, unjust, and unreasonable electricity price spike." California state utilities paid $20 billion more for energy in 2000 than in 1999 as a result, the head of the commission found.

The commission also cited the role of the Enron Corporation in the California brownouts. In June 2003 the Federal Energy Regulatory Commission (FERC) barred Enron from selling electricity and natural gas in the United States after conducting a probe into charges that Enron manipulated electricity prices during California's energy crisis. In the same month the Federal Bureau of Investigation arrested an Enron executive on charges of manipulating the price of electricity in California. Two other Enron employees, known as traders because they sold electricity, had pleaded guilty to similar charges. Despite the potential for rare widespread problems, the interconnected grid system provides necessary backup and alternate paths for power flow, resulting in much higher overall reliability than is possible with isolated systems. National or regional grids can also cope with unexpected outages such as those caused by storms, earthquakes, landslides, and forest fires, or due to human error or deliberate acts of sabotage.

POWER QUALITY

In recent years electricity has been used to power more sophisticated and technically complex manufacturing processes, computers and computer networks, and a variety of other high-technology consumer goods. These products and processes are sensitive not only to the continuity of power supply but also to the constancy of electrical frequency and voltage. Consequently, utilities are taking new measures to provide the necessary reliability and quality of electrical power, such as by providing additional electrical equipment to assure that the voltage and other characteristics of electrical power are constant.

Voltage Regulation

Long transmission lines have considerable inductance and capacitance. When a current flows through the line, inductance and capacitance have the effect of varying the voltage on the line as the current varies. Thus the supply voltage varies with the load. Several kinds of devices are used to overcome this undesirable variation in an operation called regulation of the voltage. The devices include induction regulators and three-phase synchronous motors (called synchronous condensers), both of which vary the effective amount of inductance and capacitance in the transmission circuit.

Inductance and capacitance react with a tendency to nullify one another. When a load circuit has more inductive than capacitive reactance, as almost invariably occurs in large power systems, the amount of power delivered for a given voltage and current is less than when the two are equal. The ratio of these two amounts of power is called the power factor. Because transmission-line losses are proportional to current, capacitance is added to the circuit when possible, thus bringing the power factor as nearly as possible to 1. For this reason, large capacitors are frequently inserted as a part of power-transmission systems.

World Electric Power Production

Over the period from 1950 to 2003, the most recent year for which data are available, annual world electric power production and consumption rose from slightly less than 1 trillion kilowatt-hours (kwh) to 15.9 trillion kwh. A change also took place in the type of power generation. In 1950 about two-thirds of the world's electricity came from steam-generating sources and about one-third from hydroelectric sources. In 2003 thermal sources produced 65 percent of the power, but hydropower had declined to 17 percent, and nuclear power accounted for 16 percent of the total.

Conservation

Much of the world's electricity is produced from the use of nonrenewable resources, such as natural gas, coal, oil, and uranium. Coal, oil, and natural gas contain carbon, and burning these fossil fuels contributes to global emissions of carbon dioxide and other pollutants. Scientists believe that carbon dioxide is the principal gas responsible for global warming, a steady rise in Earth's surface temperature.

Consumers of electricity can save money and help protect the environment by eliminating unnecessary use of electricity, such as turning off lights when leaving a room. Other conservation methods include buying and using energy-efficient appliances and light bulbs, and using appliances, such as washing machines and dryers, at off-peak production hours when rates are lower. Consumers may also consider environmental measures such as purchasing "green power" when it is offered by a local utility. "Green power" is usually more expensive but relies on renewable and environmentally friendly energy sources, such as wind turbines and geothermal power plants.

ELECTRIC MOTORS AND GENERATORS

Electric Motors and Generators, group of devices used to convert mechanical energy into electrical energy, or electrical energy into mechanical energy, by electromagnetic means. A machine that converts mechanical energy into electrical energy is called a generator, alternator, or dynamo, and a machine that converts electrical energy into mechanical energy is called a motor. Two related physical principles underlie the operation of generators and motors.

The first is the principle of electromagnetic induction discovered by the British scientist Michael Faraday in 1831. If a conductor is moved through a magnetic field, or if the strength of the magnetic field acting on a stationary conducting loop is made to vary, a current is set up or induced in the conductor. The converse of this principle is that of electromagnetic reaction, first observed by the French physicist Andre Marie Ampère in 1820. If a current is passed through a conductor located in a magnetic field, the field exerts a mechanical force on it.

The simplest of all dynamoelectric machines is the disk dynamo developed by Faraday. It consists of a copper disk mounted so that part of the disk, from the centre to the edge, is between the poles of a horseshoe magnet. When the disk is rotated, a current is induced between the centre of the disk and its edge by the action of the field of the magnet. The disk can be made to operate as a motor by applying a voltage between the edge of the disk and its centre, causing the disk to rotate because of the force produced by magnetic reaction.

The magnetic field of a permanent magnet is strong enough to operate only a small practical dynamo or motor. As a result, for large machines, electromagnets are employed. Both motors and generators consist of two basic units, the field, which is the electromagnet with its coils, and the armature, the structure that supports the conductors which cut the magnetic field and carry the induced current in a generator or the exciting current in a motor. The armature is usually a laminated soft-iron core around which conducting wires are wound in coils.

DIRECT-CURRENT GENERATORS

If an armature revolves between two stationary field poles, the current in the armature moves in one direction during half of each revolution and in the other direction during the other half. To produce a steady flow of unidirectional, or direct, current from such a device, it is necessary to provide a means of reversing the current flow outside the generator once during each revolution.

In older machines this reversal is accomplished by means of a commutator, a split metal ring mounted on the shaft of the armature. The two halves of the ring are insulated from each other and serve as the terminals of the armature coil. Fixed brushes of metal or carbon are held against the commutator as it revolves, connecting the coil electrically to external wires.

As the armature turns, each brush is in contact alternately with the halves of the commutator, changing position at the moment when the current in the armature coil reverses its direction. Thus there is a flow of unidirectional current in the outside circuit to which the generator is connected. DC generators are usually operated at fairly low voltages to avoid the sparking between brushes and commutator that occurs at high voltage. The highest potential commonly developed by such generators is 1500 V. In some newer machines this reversal is accomplished using power electronic devices, diode rectifiers.

Modern DC generators use drum armatures that usually consist of a large number of windings set in longitudinal slits in the armature core and connected to appropriate segments of a multiple commutator. In an armature having only one loop of wire, the current produced will rise and fall depending on the part of the magnetic field through which the loop is moving. A commutator of many segments used with a drum armature always connects the external circuit to one loop of wire moving through the high-intensity area of the field, and as a result the current delivered by the armature windings is virtually constant.

Fields of modern generators are usually equipped with four or more electromagnetic poles to increase the size and strength of the magnetic field. Sometimes smaller interpoles are added to compensate for distortions in the magnetic flux of the field caused by the magnetic effect of the armature. DC generators are commonly classified according to the method used to provide field current for energizing the field magnets. A series-wound generator has its field in series with the armature, and a shunt-wound generator has the field connected in parallel with the armature.

Compound-wound generators have part of their fields in series and part in parallel. Both shunt-wound and compound-wound generators have the advantage of delivering comparatively constant voltage under varying electrical loads. The series-wound generator is used principally to supply a constant current at variable voltage. A magneto is a small DC generator with a permanent-magnet field.

DC MOTORS

In general, DC motors are similar to DC generators in construction. They may, in fact, be described as generators "run backwards." When current is passed through the armature of a DC motor, a torque is generated by magnetic reaction, and the armature revolves.

The action of the commutator and the connections of the field coils of motors are precisely the same as those used for generators. The revolution of the armature induces a voltage in the armature windings. This induced voltage is opposite in direction to the outside voltage applied to the armature, and hence is called back voltage or counter electromotive force (emf). As the motor rotates more rapidly, the back voltage rises until it is almost equal to the applied voltage. The current is then small, and the speed of the motor will remain constant as long as the motor is not under load and is performing no mechanical work except that required to turn the armature. Under load the armature turns more slowly, reducing the back voltage and permitting a larger current to flow in the armature. The motor is thus able to receive more electric power from the source supplying it and to do more mechanical work.

Because the speed of rotation controls the flow of current in the armature, special devices must be used for starting DC motors. When the armature is at rest, it has virtually no resistance, and if the normal working voltage is applied, a large current will flow, which may damage the commutator or the armature windings. The usual means of preventing such damage is the use of a starting resistance in series with the armature to lower the current until the motor begins to develop an adequate back voltage. As the motor picks up speed, the resistance is gradually reduced, either manually or automatically.

The speed at which a DC motor operates depends on the strength of the magnetic field acting on the armature, as well as on the armature current. The stronger the field, the slower is the rate of rotation needed to generate a back voltage large enough to counteract the applied voltage. For this reason the speed of DC motors can be controlled by varying the field current.

ALTERNATING-CURRENT GENERATORS

A simple generator without a commutator will produce an electric current that alternates in direction as the armature revolves. Such alternating current is advantageous for electric power transmission, and hence most large electric generators are of the AC type. In its simplest form, an AC generator differs from a DC generator in only two particulars: the ends of its armature winding are brought out to solid unsegmented slip rings on the generator shaft instead of to commutators, and the field coils are energized by an external DC source rather than by the generator itself.

Low-speed AC generators are built with as many as 100 poles, both to improve their efficiency and to attain more easily the frequency desired. Alternators driven by high-speed turbines, however, are often two-pole machines. The frequency of the current delivered by an AC generator is equal to half the product of the number of poles and the number of revolutions per second of the armature.

It is often desirable to generate as high a voltage as possible, and rotating armatures are not practical in such applications because of the possibility of sparking between brushes and slip rings and the danger of mechanical failures that might cause short circuits.

Alternators are therefore constructed with a stationary armature within which revolves a rotor composed of a number of field magnets. The principle of operation is exactly the same as that of the AC generator described, except that the magnetic field (rather than the conductors of the armature) is in motion.

The current generated by the alternators described above rises to a peak, sinks to zero, drops to a negative peak, and rises again to zero a number of times each second, depending on the frequency for which the machine is designed. Such current is known as single-phase alternating current. If, however, the armature is composed of two windings, mounted at right angles to each other, and provided with separate external connections, two current waves will be produced, each of which will be at its maximum when the other is at zero. Such current is called two-phase alternating current. If three armature windings are set at 120° to each other, current will be produced in the form of a triple wave, known as three-phase alternating current. A larger number of phases may be obtained by increasing the number of windings in the armature, but in modern electrical-engineering practice three-phase alternating current is most commonly used, and the three-phase alternator is the dynamoelectric machine typically employed for the generation of electric power. Voltages as high as 13,200 are common in alternators.

AC Motors

Two basic types of motors are designed to operate on polyphase alternating current, synchronous motors and induction motors. The synchronous motor is essentially a three-phase alternator operated in reverse. The field magnets are mounted on the rotor and are excited by direct current, and the armature winding is divided into three parts and fed with three-phase alternating current.

The variation of the three waves of current in the armature causes a varying magnetic reaction with the poles of the field magnets, and makes the field rotate at a constant speed that is determined by the frequency of the current in the AC power line. The constant speed of a synchronous motor is advantageous in certain devices; however, in applications where the mechanical load on the motor becomes very great, synchronous motors cannot be used, because if the motor slows down under load it will "fall out of step" with the frequency of the current and come to a stop. Synchronous motors can be made to operate from a single-phase power source by the inclusion of suitable circuit elements that cause a rotating magnetic field.

The simplest of all electric motors is the squirrel-cage type of induction motor used with a three-phase supply. The stator, or stationary armature, of the squirrel-cage motor consists of three fixed coils similar to the armature of the synchronous motor. The rotating member consists of a core in which are imbedded a series of heavy conductors arranged in a circle around the shaft and parallel to it. With the core removed, the rotor conductors resemble in form the cylindrical cages once used to exercise pet squirrels.

The three-phase current flowing in the stator windings generates a rotating magnetic field. This field induces a current in the conductors of the cage. The magnetic reaction between the rotating field and the current-carrying conductors of the rotor makes the rotor turn. If the rotor is revolving at exactly the same speed as the magnetic field, no currents will be induced in it, and hence the rotor should not turn at a synchronous speed. In operation the speeds of rotation of the rotor and the field differ by about 2 to 5 percent. This speed difference is known as slip. Motors with squirrel-cage rotors can be used on single-phase alternating current by means of various arrangements of inductance and capacitance that alter the characteristics of the single-phase voltage and make it resemble a two-phase voltage. Such motors are called split-phase motors or condenser motors (or capacitor motors), depending on the arrangement used. Single-phase squirrel-cage motors do not have a large starting torque, and for applications where such torque is required, repulsion-induction motors are used.

A repulsion-induction motor may be of the split-phase or condenser type, but has a manual or automatic switch that allows current to flow between brushes on the commutator when the motor is starting, and short-circuits all commutator segments after the motor reaches a critical speed. Repulsion-induction motors are so named because their starting torque depends on the repulsion between the rotor and the stator, and their torque while running depends on induction. Series-wound motors with commutators, which will operate on direct or alternating current, are called universal motors. They are usually made only in small sizes and are commonly used in household appliances.

MISCELLANEOUS MACHINES

For special applications several combined types of dynamoelectric machines are employed. It is frequently desirable to change from direct to alternating current or vice versa, or to change the voltage of a DC supply, or the frequency or phase of an AC supply. One means of accomplishing such changes is to use a motor operating from the available type of electric supply to drive a generator delivering the current and voltage wanted. Motor generators, consisting of an appropriate motor mechanically coupled to an appropriate generator, can accomplish most of the indicated conversions.

A rotary converter is a machine for converting alternating to direct current, using separate windings on a common rotating armature. The AC supply voltage is applied to the armature through slip rings, and the DC voltage is led out of the machine through a separate commutator. A dynamotor, which is usually used to convert low-voltage direct current to high-voltage direct current, is a similar machine that has separate armature windings.

Pairs of machines known as synchros, selsyns, or autosyns are used to transmit torque or mechanical movement from one place to another by electrical means. They consist of pairs of motors with stationary fields and armatures wound with three sets of coils similar to those of a three-phase alternator. In use, the armatures of selsyns are connected electrically in parallel to each other but not to any external source.

The field coils are connected in parallel to an external AC source. When the armatures of both selsyns are in the same position relative to the magnetic fields of their respective machines, the currents induced in the armature coils will be equal and will cancel each other out. When one of the armatures is moved, however, an imbalance is created that will cause a current to be induced in the other armature. The magnetic reaction to this current will move the second armature until it is in the same relative position as the first. Selsyns are widely used for remote-control and remote-indicating instruments where it is inconvenient or impossible to make a mechanical connection.

DC machines known as amplidynes or rotortrols, which have several field windings, may be used as power amplifiers. A small change in the power supplied to one field winding produces a much larger corresponding change in the power output of the machine. These electrodynamic amplifiers are frequently employed in servomechanism and other control systems.

Chapter 8

Wind Energy

ENERGY FROM WIND

Wind is simple air in motion. It is caused by the uneven heating of the earth's surface by the sun. Since the earth's surface is made of very different types of land and water, it absorbs the sun's heat at different rates. The land heats up more quickly than the air over water. The warm air over the land expands and rises, and the heavier, cooler air rushes in to take its place, creating winds.

At night, the winds are reversed because the air cools more rapidly over land than over water. In the same way, the large atmospheric winds that circle the earth are created because the land near the earth's equator is heated more by the sun than the land near the North and South Poles. Today, wind energy is mainly used to generate electricity. Wind is called a renewable energy source because the wind will blow as long as the sun shines.

Wind Energy, energy contained in the force of the winds blowing across the earth's surface. When harnessed, wind energy can be converted into mechanical energy for performing work such as pumping water, grinding grain, and milling lumber. By connecting a spinning rotor (an assembly of blades attached to a hub) to an electric generator, modern wind turbines convert wind energy, which turns the rotor, into electrical energy.

Wind is created when air that has been warmed over sun-heated land rises, leaving a vacuum in the space it once occupied. Cooler surrounding air then rushes in to fill the vacuum. This movement of rushing air is what we know as wind.Egyptians may have been the first to capture wind energy when they sailed boats up the Nile River beginning around the 4th century.

For centuries afterward, wind-powered sailing vessels plied the world's seas and oceans, serving as the principal form of commercial transport. Wind energy has been harnessed on land since the first windmill was developed by the ancient Persians in the 7th century.

Windmills have since been used to mill grain, pump water, saw timber, and provide other forms of mechanical energy.Because wind is a clean and renewable source of energy, modern wind turbines had been installed in 26 countries, including such nations as Germany, Denmark, India, China, and the United States, to supplement more traditional sources of electric power, such as burning coal.

Design improvements such as more efficient rotor blades combined with an increase in the numbers of wind turbines installed, have helped increase the world's wind energy generating capacity by nearly 150 percent since 1990. In 2006 the United States became the world's third largest producer of energy from wind power, generating more than 11.5 megawatts of electricity.

WINDMILL

Windmill, machine that converts wind into useful energy. This energy is derived from the force of wind acting on oblique blades or sails that radiate from a shaft. The turning shaft may be connected to machinery used to perform such work as milling grain, pumping water, or generating electricity. When the shaft is connected to a load, such as a pump, the device is typically called a windmill. When it is used to generate electricity, it is known as a wind turbine generator.

Uses and Improvements

Besides milling grain and irrigating farmland, windmills developed from the 15th century to the 19th century were adapted to a variety of tasks, including pumping seawater from land below sea level, sawing wood, making paper, pressing oil from seeds, and grinding many different materials. By the 19th century the Dutch had built about 9000 windmills.Of the major improvements on the windmill, the most important was the fantail, a mechanism invented in 1745 that automatically rotates the sails into the wind. In 1772 the spring sail was developed.

This type of sail consists of wood shutters, the openings of which can be controlled either manually or automatically to maintain a constant sail speed in winds of varying speeds. Other improvements include air brakes to stop the sails from rotating and the use of propellerlike airfoils in place of sails, which increases the usefulness of mills in light winds.Water-pumping windmills were widely employed during the settlement of the western United States.

The use of wind turbines for generating electricity was pioneered in Denmark late in the 1890s. Small wind turbine generators supplied electricity to many rural communities in the United States until the 1930s, when power lines were extended across the nation. Large wind turbines were also built during this time. The largest was the Smith-Putnam generator, installed in 1941 at Grandpa's Knob, near Rutland, Vermont.

Modern Wind Turbines

Modern wind turbines are propelled by one of two effects: drag, by which wind pushes the blades; and lift, by which the blades are moved in the same way an airplane's wing rises on an air current. Turbines operated by lift turn more rapidly and are inherently more efficient. Wind turbines can be classified as horizontal-axis machines, with their main shafts parallel to the ground, or vertical-axis machines, with shafts perpendicular to the ground.

Horizontal-axis turbines used to generate electricity have one to three blades; those used for pumping may have many more. The most common vertical-axis machines, named after their designers, are the Savonius, used primarily for pumping, and the Darrieus, a higher-speed machine resembling an eggbeater.

WIND MACHINES WORK

Like old fashioned windmills, today's wind machines use blades to collect the wind's kinetic energy. Windmills work because they slow down the speed of the wind. The wind flows over the airfoil shaped blades causing lift, like the effect on airplane wings, causing them to turn. The blades are connected to a drive shaft that turns an electric generator to produce electricity. With the new wind machines, there is still the problem of what to do when the wind isn't blowing. At those times, other types of power plants must be used to make electricity.

Potential Turbine Power

The amount of power transferred to a wind turbine is directly proportional to the density of the air, the area swept out by the rotor, and the cube of the wind speed.

The usable power P available in the wind is given by:

$$P = \frac{1}{2}\alpha\rho\pi r^2 v^2 ,$$

where P = power in watts, α = an efficiency factor determined by the design of the turbine, ρ = mass density of air in kilograms per cubic meter, r = radius of the wind turbine in meters, and v = velocity of the air in meters per second.

As the wind turbine extracts energy from the air flow, the air is slowed down, which causes it to spread out. Albert Betz, a German physicist, determined in early 1900s that a wind turbine can extract at most 59% of the energy that would otherwise flow through the turbine's cross section, that is *á* can never be higher than 0.59 in the above equation. The Betz limit applies regardless of the design of the turbine.

This equation shows the effects of the mass rate of flow of air traveling through the turbine, and the energy of each unit mass of air flow due to its velocity. As an example, on a cool 15 °C (59 °F) day at sea level, air density is 1.225 kilograms per cubic metre. An 8 m/s (28.8 km/h or 18 mi/h) breeze blowing through a 100 meter diameter rotor would move almost 77,000 kilograms of air per second through the swept area. The total power of the example breeze through a

100 meter diameter rotor would be about 2.5 megawatts. Betz' law states that no more than 1.5 megawatts could be extracted.

TYPES OF WIND MACHINES

There are two types of wind machines (turbines) used today based on the direction of the rotating shaft (axis): horizontal–axis wind machines and vertical-axis wind machines. The size of wind machines varies widely. Small turbines used to power a single home or business may have a capacity of less than 100 kilowatts. Some large commercial sized turbines may have a capacity of 5 million watts, or 5 megawatts. Larger turbines are often grouped together into wind farms that provide power to the electrical grid.

Horizontal-axis

Most wind machines being used today are the horizontal-axis type. Horizontal-axis wind machines have blades like airplane propellers. A typical horizontal wind machine stands as tall as a 20-story building and has three blades that span 200 feet across. The largest wind machines in the world have blades longer than a football field! Wind machines stand tall and wide to capture more wind.

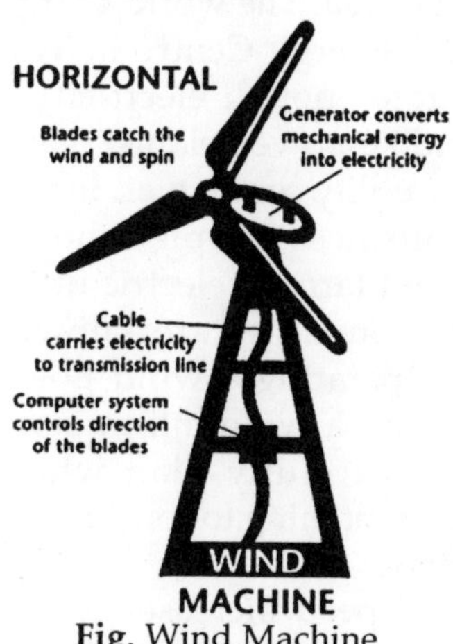

Fig. Wind Machine

Vertical-axis

Vertical–axis wind machines have blades that go from top

to bottom and the most common type looks like a giant two-bladed egg beaters. The type of vertical wind machine typically stands 100 feet tall and 50 feet wide. Vertical-axis wind machines make up only a very small percent of the wind machines used today.

The Wind Amplified Rotor Platform (WARP) is a different kind of wind system that is designed to be more efficient and use less land than wind machines in use today. The WARP does not use large blades; instead, it looks like a stack of wheel rims. Each module has a pair of small, high capacity turbines mounted to both of its concave wind amplifier module channel surfaces. The concave surfaces channel wind toward the turbines, amplifying wind speeds by 50 percent or more. Eneco, the company that designed WARP, plans to market the technology to power offshore oil platforms and wireless telecommunications systems.

WIND POWER PLANTS

Wind power plants, or wind farms as they are sometimes called, are clusters of wind machines used to produce electricity. A wind farm usually has dozens of wind machines scattered over a large area. The world's largest wind farm, the Horse Hollow Wind Energy Centre in Texas, has 421 wind turbines that generate enough electricity to power 230,000 homes per year. Unlike power plants, many wind plants are not owned by public utility companies. Instead they are owned and operated by business people who sell the electricity produced on the wind farm to electric utilities.

These private companies are known as Independent Power Producers. Operating a wind power plant is not as simple as just building a windmill in a windy place. Wind plant owners must carefully plan where to locate their machines. One important thing to consider is how fast and how much the wind blows.

As a rule, wind speed increases with altitude and over open areas with no windbreaks. Good sites for wind plants are the tops of smooth, rounded hills, open plains or shorelines, and mountain gaps that produce wind funneling. Wind speed varies throughout the country. It also varies from season to season. In Tehachapi, California, the wind blows more from April through October than it does in the winter. This is because of the extreme heating of the Mojave Desert during the summer months. The hot air over the desert rises, and the cooler, denser air above the Pacific Ocean rushes through the Tehachapi mountain pass to take its place.

In a state like Montana, on the other hand, the wind blows more during the winter. Fortunately, these seasonal variations are a good match for the electricity demands of the regions. In California, people use more electricity during the summer for air conditioners. In Montana, people use more electricity during the winter months for he ating.

WIND PRODUCTION

Wind machines generated a total of 17.8 billion kWh per year of electricity, enough to serve more than 1.6 million households. This is enough electricity to power a city the size of Chicago, but it is only a small fraction of the nation's total electricity production, about 0.4 percent. The amount of electricity generated from wind has been growing fast in recent years, tripling since 1998.

New technologies have decreased the cost of producing electricity from wind, and growth in wind power has been encouraged by tax breaks for renewable energy and green pricing programs. Many utilities around the country offer green pricing options that allow customers the choice to pay more for electricity that comes from renewable sources. Wind machines generate electricity in 25 different states in 2005. The states with the most wind production are California, Texas, Iowa, Minnesota, and Oklahoma. The United States ranks third in the world in wind power capacity, behind Germany and Spain and before India. Denmark ranks number five in the world in wind power capacity but generates 20 percent of its electricity from wind. Most of the wind power plants in the world are located in Europe and in the United States where government programs have helped support wind power development.

WIND AND THE ENVIRONMENT

Oil shortages pushed the development of alternative energy sources. In the 1990s, the push came from a renewed concern for the environment in response to scientific studies indicating potential changes to the global climate if the use of fossil fuels continues to increase. Wind energy is an economical power resource in many areas of the country. Wind is a clean fuel; wind farms produce no air or water pollution because no fuel is burned. Growing concern about emissions from fossil fuel generation, increased government support, and higher costs for fossil fuels (especially natural gas and coal) have helped wind power capacity in the United States grow substantially over the last 10 years. The most serious environmental drawbacks to wind machines may be their negative effect on wild bird populations and the visual impact on the landscape. To some, the glistening blades of windmills on the horizon are an eyesore; to others, they're a beautiful alternative to conventional power plants.

BETZ' LAW

Betz' law reflects a theory for flow machines, developed by Albert Betz. It shows the maximum possible energy that may be derived by means of an infinitely thin rotor from a fluid flowing at a certain speed.

In order to calculate the maximum theoretical efficiency of a thin rotor one imagines it to be replaced by a disc that withdraws energy from the fluid passing through it. At a certain distance behind this disc, the fluid, that has passed through, flows with a reduced velocity.

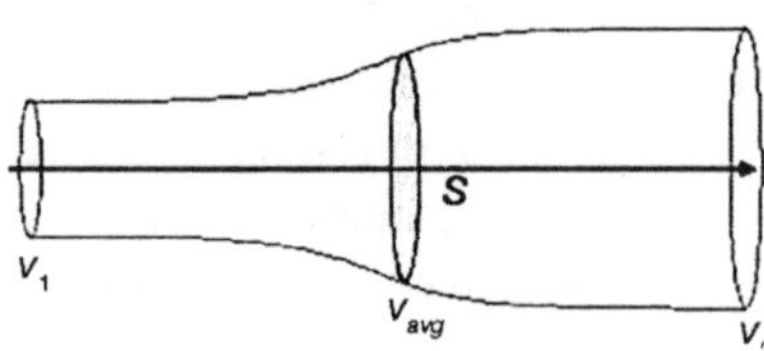

Fig. Fluid Flow Through a Disk-shaped Actuator.

Let v_1 be the speed of the fluid in front of the rotor and v_2 that of the fluid downstream from it. The mean flow velocity through the disc representing the rotor is v_{avg}, where

$$v_{avg} = \frac{1}{2}\cdot(v_1 + v_2)$$

With the area of the disc equal to S, and with ρ = fluid density, the mass flow rate (the mass of fluid flowing per unit time) is given by:

$$\dot{m} = \rho\cdot S\cdot v_{avg} = \frac{\rho\cdot S\cdot(v_1+v_2)}{2}$$

The power delivered is the difference between the kinetic energies of the flows approaching and leaving the rotor in unit time:

$$\begin{aligned}\dot{E} &= \frac{1}{2}\cdot\dot{m}\cdot\left(v_1^2 - v_2^2\right)\\ &= \frac{1}{4}\cdot\rho\cdot S\cdot(v_1+v_2)\cdot\left(v_1^2 - v_2^2\right)\\ &= \frac{1}{4}\cdot\rho\cdot S\cdot v_1^3\cdot\left(1-\left(\frac{v_2}{v_1}\right)^2+\left(\frac{v_2}{v_1}\right)-\left(\frac{v_2}{v_1}\right)^3\right)\end{aligned}$$

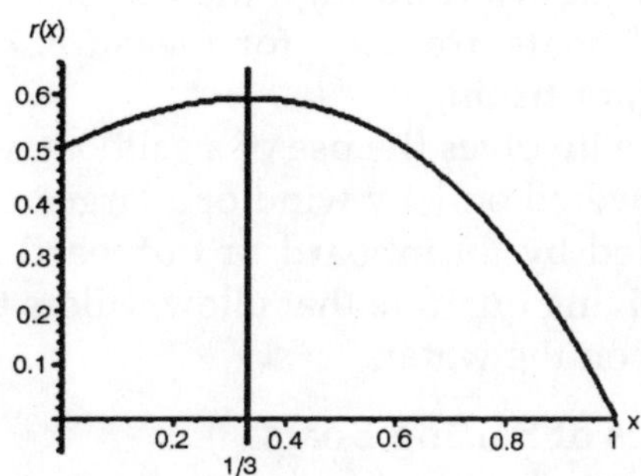

The horizontal axis reflects the ratio $\frac{v_2}{v_1}$, the vertical axis is the "coefficient of performance" C_p.

By differentiating $\dot{E}$ with respect to $\frac{v_2}{v_1}$ for a given fluid speed v_1 and a given area S one finds the *maximum* or *minimum* value for $\dot{E}$. The result is that $\dot{E}$ reaches maximum value when $\frac{v_2}{v_1} = \frac{1}{3}$.

Substituting this value results in:

$$P_{max} = \frac{16}{27} \cdot \frac{1}{2} \cdot \rho \cdot S \cdot v_1^3 .$$

The work rate obtainable from a cylinder of fluid with area S and velocity v_1 is:.

The "coefficient of performance" C_p ($= \frac{P}{P_{max}}$) has a maximum value of: $C_{p.max} = \frac{16}{27} = 0.593$ (or 59.3%; however, coefficients of performance are usually expressed as a decimal, not a percentage).Rotor losses are the most significant energy losses in, a wind mill. It is, therefore, important to reduce these as much as possible. Modern rotors achieve values for C_p in the range of 0.4 to 0.5, which is 70 to 80% of the theoretically possible.

SAILING

Sailing, propulsion of a boat or ship by means of the driving force of the wind through the use of sails. In sailing, noncommercial boats are used for pleasure, especially for cruising, racing, or fishing.

The pastime involves the use of a sailboat, which may be a small boat powered only by wind or a larger vessel that can also be propelled by an inboard or outboard motor. Some sailboats have living quarters that allow sailors to spend long periods of time on the water.

Basic Principles of Sailing Boats

Methods of sailing vary according to the manner in which boats are rigged, but the essential principles of sailing are the same for all craft.

The simplest and most easily understood point of sailing is called in nautical terms sailing before the wind. The term running before the wind is also used. As the term indicates, the boat follows the same course that the wind is blowing. The sail or sails are set at approximately a 90° angle to the longitudinal axis of the boat, with power derived from the push of the wind on the sails' back surfaces.

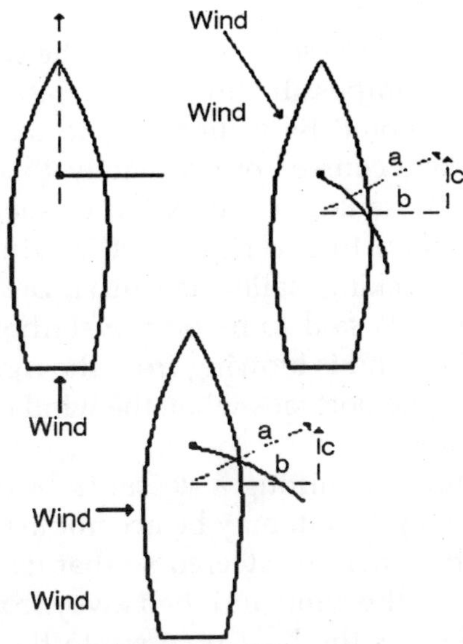

Fig. Sails of Boat.

In sailing off the wind, as shown in the middle diagram, the wind reaches the craft from the side, or beam, and the sails are set at approximately 45° from the axis of the craft. In this sailing position, the wind exerts a pulling rather than a pushing action on the sails, which act as airfoils, like the wings of an airplane.

The general principle of wind action is that the wind flows at a greater rate of speed along the forward surface of the sail, creating an area of lower pressure ahead of the sail. The actual force exerted by the wind is at right angles to the sail, as indicated by the dotted line *a*.

This force would tend to drive the boat at an oblique angle if the hull of the boat were perfectly flat. Every sailboat, however, is equipped with a fixed keel or a retractable centerboard, which acts as a flat longitudinal plane to prevent the boat from moving sideways through the water.

The effect of this plane is shown by the dotted line b, and the actual course of the boat, the result of both the force of the wind and the resisting force of the keel, is the dotted line *c*, representing forward motion.

TACKING

If boats were able to sail only before the wind and off the wind, it would be impossible to reach a destination upwind from the starting point. By sailing on the wind, however, a sailboat can make a course approximately 45° away from the wind direction. By sailing a succession of such courses, first to the left and then to the right of the wind direction, a maneuver called tacking, sailboats can zigzag in an upwind direction. A vessel is said to be on the starboard tack when sailing so that the wind is blowing from the right or starboard side, and to be on the port tack when the wind is blowing from the left or port side.

The procedure of shifting a vessel from one tack to the other, called coming about, may be accomplished in either of two ways. The boat may be steered so that its bow (the front end) points up into the wind and then away from the wind on the opposite tack. As the boat points into the wind, it loses speed, the sails being pressed directly backward by the wind. Then as the bow moves away from the wind on the other tack, the sails fill with wind again and assume a position on the other side of the vessel.

During the time of coming about, the boat is receiving no motive force from the wind; it must rely on its momentum to maintain enough speed so that it can be steered onto the opposite tack. When the boat does not have sufficient momentum and stops with its bow pointing into the wind and its sails useless, it is said to be in irons.

The other method of changing tack consists of steering the boat away from the direction of the wind, until the wind fills the sails from the other side and the boat is on the other tack. In fore-and-aft-rigged vessels, this maneuver is called *jibing* or *gybing,* and in square-rigged ships it is known as *wearing*. When running before the wind, a slight shift of wind may cause a boat to jibe unintentionally.

Such jibing is dangerous because of the speed with which the heavy booms, or spars, at the foot of the sails sweep across the decks of the vessel from one side to the other, and also because of the danger of breaking spars. In wild jibing, control can be lost momentarily and, if the seas are high, a small boat can *broach*—that is, veer on its side with danger of swamping or capsizing. An unintentional jibe in a heavy wind frequently has enough force to break the masts of a vessel. When jibing intentionally, careful sailors always haul in on the boom while turning, so that the boom will travel only a short distance when the wind reaches the other side of the sails.

Reefing

During stormy weather, the area of sail exposed can be reduced by another procedure of sail control known as reefing. Reefing is accomplished by bunching up a portion of the slackened sail along the yard or the boom and then securing the folded canvas with small ropes called *reef points*. The part of the sail thus taken in is called the reef.

On all sailing ships, sails are hauled up and, to some degree, controlled in accordance with wind direction by ropes called *sheets* and *halyards*. For the most important of these ropes, as well as other portions of the fittings and rigging of sailing ships.

Craft

Boats using sails for propulsive power may be classified as sailing cruisers, day sailers, auxiliary cruisers, and motor sailers. As the names suggest, both the sailing cruiser and the day sailer are driven solely by sails. The sailing cruiser is longer and beamier (broader) than the day sailer and, unlike the latter, possesses living facilities. An auxiliary cruiser is a sailing cruiser equipped with an inboard engine. A motor sailer is an underrigged, heavily powered auxiliary cruiser—that is, a vessel dependent primarily on its engine or engines, but capable of maintaining headway under sail. The cabin cruiser, which is equipped with complete living quarters for two or more persons, is the most popular type of habitable motor-powered craft used in yachting.

The larger craft are powered by one to three inboard gasoline or diesel engines. Many of the smaller types, including those craft that can be assembled from kits, are driven by one or two outboard motors.

Sailing craft used for racing may be grouped into three main categories: one-design, rating, and handicap. One-design boats come in numerous classes, and all boats belonging to a particular class are identical. In one-design racing, consequently, success is determined by seamanship rather than by differences in design or equipment. Especially popular with sailors of modest means are the smaller one-design boats, of the Sunfish or the Laser class.

These range from about 2 to 12 m (about 7 to 40 ft) in length. Rating-class boats differ slightly from one another in certain particulars such as length of hull, displacement, and sail area. All boats of a given class conform, however, to a certain overall rating arrived at in accordance with a set mathematical formula. The success of a rating-class sailboat consequently depends to some extent on the expertise of its designer. Boats differing widely in size and design compete in handicap racing. The boats are measured according to certain rating criteria and are assigned appropriate time allowances. The handicap system enables small and shallow-draft boats to race on equal terms with larger and deeper craft.

ADVANTAGES AND DISADVANTAGES OF WIND ENERGY

The conversion of wind energy to various other useful forms, like electricity, is known as wind power. Wind energy is converted into these forms using wind turbines. Wind energy can be converted into electrical energy by the use of a electrical generator.

The first use of wind energy was through wind mills. Wind mills had engines which used to produce energy using wind. This energy was usually used in rural and agricultural areas for grinding, pumping, hammering and various farm needs. Even today, wind energy is used in large scale wind farms to provide electricity to rural areas and other far reaching locations.

Advantages

One of the greatest advantages of Wind Energy is that it

is ample. Secondly, wind energy is renewable. Some other advantages of Wind Energy are that it is widely distributed, cheap, and also reducing toxic gas emissions. Wind Energy is also advantageous over traditional methods of creating energy, in the sense that it is getting cheaper and cheaper to produce wind energy. Wind Energy may soon be the cheapest way to produce energy on a large scale.

The cost of producing wind energy has come down by at least eighty percent since the eighties. Along with economy, Wind Energy is also said to diminish the greenhouse effect. Also, wind energy generates no pollution. Wind Energy is also a more permanent type of energy. The wind will exist till the time the sun exists, which is roughly another four billion years. Theoretically, if all the wind power available to humankind is harnessed, there can be ten times of energy we use, readily available.

One other advantage of wind energy that it is readily available around the globe, and therefore there would be no need of dependence for energy for any country. Wind energy may be the answer to the globe's question of energy in the face of the rising petroleum and gas prices.

Disadvantages

However, there are some disadvantages for wind energy, which may put a dampener in its popularity. Though the costs of creating wind energy is going down, even today a large number of turbines have to be built to generate a proper amount of wind energy. Though wind power is non-polluting, the turbines may create a lot of noise, which indirectly contributes to noise pollution.

Wind can never be predicted.Even the most advanced machinery may come out a cropper while predicting weather and wind conditions. Since wind energy will require knowledge of the weather and wind conditions on long term basis, it may be a bit impractical.

Therefore, in areas where a large amount of wind energy is needed, one cannot depend completely on wind. Many potential wind farms, places where wind energy can be produced on a large scale, are far away from places for which wind energy is best suited. Therefore, the economical nature of wind energy may take a beating in terms of costs of new substations and transmission lines.

Wind Energy is non-dispatchable. This may also put a spanner in depending upon wind power as a primary energy supplier. Wind energy depends upon the wind in an area and therefore is a variable source of energy. The amount of wind supplied to a place and the amount of energy produced from it will depend on various factors like wind speeds and the turbine characteristics. Some critics also wonder whether wind energy can be used in areas of high demand.

WIND POWER

Sun's rays heat the atmosphere of the earth in different degrees. There are parts of the earth that get more heat, consequently the air in those parts also gets warm and rises whereas in the other places which do not get enough heat, the air remains cold and remains close to the surface. But that one section of air, which is heated and rises, creates a void and the cold air blows to fill in the vacuum. This creates the movement of air and that is what we called wind. Now when this wind is used to get power it is called wind power. Wind power was used in the earlier times as well with the means of windmills though to get mechanical job done and not to produce electrical energy.

Method of Wind power Production

The energy of wind is converted to wind power with the help of turbines, which rotate as the wind blows and the rotation of the turbine produce power, which is distributed through the grid that is connected to the turbine.

A turbine in average utilizes approximately 59% of the wind that passes through it but the placement of turbine is very important. It has to be kept in mind that there should be scope for transmission lines and the environmental factors since these would have a direct impact on the whole project of power generation. These are the important pointers apart from wind availability and cost of land, which are to be considered.

Advantages and Disadvantages of Wind Power

As stated earlier Wind power generation does not involve pollution and this power source is also renewable, but apart from these there are some more advantages of Wind Power, which are as follows:

- Wind is easily available in all parts of the world and can become a cheap source of power for developing countries that face the crisis of power on large scale.
- It is nature's endowment therefore is available free of cost.
- Wind Turbines do not consume additional power for power generation.
- The turbines do not take much space and therefore are very convenient to install. They are also available in various models and potentials. As their potentials differ they are useful for all kinds of industries.
- The disadvantages of Wind Power are few in comparison to the advantages. The one disadvantage that makes it a less used and reliable source of energy is its varying strength. It is not same all the time and therefore the amount of power generated can vary.
- The turbines at times produce a lot of noise and there have been objections to the fact that the wind farms are being established in the countryside which is spoiling the scenic beauty.

There are advantages as well as disadvantages but it is better to improve the technology and get away with all the disadvantages because that would be a wise move towards a cleaner atmosphere.

BISON AND WIND POWER

Bison and wind power offer us strikingly parallel opportunities — ways to utilize the endlessly renewable resources of grass and wind that can preserve the Plains ecosystem and support its human inhabitants indefinitely.

They are thus not at all the unrelated subjects they might at first seem. In the great test case that the Plains present to us, if we learn to honour and use the bison and the winds we will have made a fundamental transition in attitude from reckless exploitation to respectful coexistence with the great natural forces of the planet. Both bison and wind are there, waiting to teach us the imperative lessons of sustainability.

Only two generations ago, windmills were the normal power source for wells throughout agricultural America. Windmill, water-storage tower, and house constituted the basic image of the isolated farmstead, imprinted on our minds through a hundred Western movies.

Spinning day and night, the faithful windmills pumped up life-giving water, the creak of their gears providing reassurance that all was well. Even today, amazingly durable old-fashioned windmills provide cheap and reliable pumping to many farms and ranches.

Finding and rebuilding an old windmill, in fact, has now become the pride of former city dwellers moving to the country — a sign of nostalgia for a time when it was universal for farmers to draw energy directly from sun and wind and live sustainably on the land with their animals. But wind power is poised to return to the Plains in a new, highly sophisticated, modern-technology form.

As in any region, people on the Plains use energy for many purposes. Transportation, heating of dwellings and other buildings, refrigeration and air-conditioning, lighting, and water heating are the biggest direct ways in which energy is consumed — most of it in the forms of gasoline, electricity, and natural gas.

Plains coal is used in large quantities to produce electricity. But overall, the Plains depend on (as does the United States as a whole) imported fossil-fuel energy — a regime that cannot be maintained indefinitely.

A basic criterion for a society's sustainability lies in its material and energy "throughputs" everything that passes through the industrial system. Mining and smelting of metals, casting and stamping, refining of oil and molding of plastics, painting and finishing all the processes of manufacturing cause substantial environmental impacts, as do the combustion-driven shipping and distribution of products. Such are the scale and intensity of contemporary industrial activity that these impacts are overwhelming the capacity of the earth's natural systems to absorb them.

Thus, either a scale-down of industrial production will have to be accomplished voluntarily or nature will enforce it. For instance, executives of major global insurance companies are beginning to think that increased floods and other weather-connected disasters may be due to carbon dioxide-driven global warming. Higher disaster risks translate directly into higher insurance costs, which feed back to reduce risk-exposed human activities.

Similarly, energy-intensive farming and ranching convert imported fossil-fuel energy (turned into fertilizer, herbicides and pesticides, equipment and its fuel supplies, animal feed, and so on) into beef and wheat.

The unsustainability of this situation is increasingly manifested through the rising operation costs experienced by farmers and ranchers.

On the Plains, the transition from fossil fuels to renewable energy (solar, both thermal and photovoltaic; wind power; and existing hydroelectric development) will actually be easier than in other regions because of the Plains' sparse population and rich resources of grass and wind. Whereas solar installations will probably not become widespread until the middle of the next century, wind power is on the verge of wide deployment now.

It is sometimes argued that the transition to a sustainable-energy system, on the Plains and elsewhere, can be postponed indefinitely or avoided through reliance on coal and natural gas. The discovery of new large reserves of gas, in particular, has been welcomed by some as a sign that fossil fuels can power a century of continued economic growth.

New supplies of carbon fuels, however, are in fact a curse rather than a blessing, since they will encourage the already industrialized nations (and rapidly developing larger nations like China and India) to avoid adopting renewable energy systems. However, burning up the planet's fossil fuel reserves will almost certainly exacerbate global warming to disastrous levels. Though gas produces somewhat less pollution and carbon dioxide than does coal, it can only serve as a transitional fuel to the renewable era.

A few scientists have begun to suspect, incidentally, that natural gas is a geological product still being created deep within the earth and not, like coal and oil, a fossilized biological product. Even if this turns out to be true, the long-term picture does not change; the gas-formation process, if it exists, must be enormously slower than human use of gas. Gas can serve, to some extent, only as a "bridge fuel" between the fossil-fuel era and a renewable-energy future.

A Plains Resource

Generated by differences in solar heat distribution over different regions of the planet, winds blow most strongly where they are unobstructed by trees, buildings, or rough terrain. Thus, the strongest and most constant winds are found at sea, along seacoasts, and in relatively flat country like the Plains, which are sometimes called the "sea of grass." The energy available from wind is proportional to the cube of wind speed. That is, wind three times as fast has twenty-seven times the energy, so wind installations are most efficient in high-wind locations. Many places qualify on the Plains.

Bison have always lived in harmony with the wind, and they survive winter weather so cold that ice forms on their beards. They face into blizzards, not away from them as cattle do, so they never crowd up against fences and freeze to death. When winter winds thin the snow on a rise, bison go there to nose down and find the underlying grass.

Wind and bison have fit together in the ecological past, and they have a sustainable future together as well. They are both dominant features of the original Plains landscape, and we can learn to rely on them both.

Our daily experience does not equip us to grasp the massive energy potential of the winds. But North Dakota's wind alone could provide 36 percent of today's total national electric demand; the state has more wind-energy potential than California, heretofore the nation's leader in wind power. The Great Plains as a whole could meet the nation's energy needs many times over.

And wind installations, including the narrow service roads they need, occupy only a fraction of the land's surface, leaving plenty of room for bison. Well-designed "wind-farm" roads and towers do not cause soil erosion; wind-farm transmission lines can and should be placed underground wherever scenic values are important. And an important economic appeal of wind power is that leasing land for wind farms can provide much-needed supplementary income to ranchers, helping to preserve open space that might otherwise be developed into condominiums and strip malls.

WORLD WIND POWER TODAY

The Plains need not play a pioneering role in wind power; they can merely follow along behind striking developments elsewhere. A wind farm may soon replace one of the reactors at the Chernobyl nuclear plant, and there are twenty thousand wind turbines spinning worldwide. Northern European wind-power installations are expanding rapidly, but projects are also under way in Argentina, China, India, Mexico, New Zealand, Spain, and other countries.

In New England, the United Kingdom, and Poland, among other places, wind turbines producing large amounts of energy are being planned for mounting, like oil-well towers, in shallow offshore waters. Some energy experts believe that wind power will come to supply 20 percent of the world's energy.

In the past, public and media perceptions of wind power, along with the decisions of U.S. utility executives, were colored by a series of spectacular failed experiments with giant wind machines by Boeing, Pacific Gas and Electric Company, and the U.S. Department of Energy. The blades of some of these early monsters were as long as football fields, and, predictably enough, they vibrated dangerously.

But the learning curve in developing more modest-sized and easily managed machines has been steep. Three or four technical generations of design experience in California and, more recently, in Europe have led to much more efficient blades and to more responsive, electronically controlled turbines.

Wind is now a thoroughly proven technology. Moreover, the maintenance and repair of wind installations is a nondemanding, medium-tech business for which the Plains could easily supply the work force.

Working on wind machines does not involve the heroic precautions for radiation safety or drastic interruptions to service that nuclear plants are subject to; when a wind turbine needs repairs, hundreds of others nearby keep right on working. In certain damp locations, like Vermont's Green Mountains, wind machines are subject to occasional icing up in winter. But otherwise wind has become a reliable, almost humdrum source of power. It will be right at home in the American heartland, along with bison.

But the peaks and valleys of wind generation will soon be smoothable by the deployment of a new generation of storage devices, including high-tech flywheels as well as new types of batteries.

Sophisticated flywheels that store impressive amounts of rotational energy are a still-unfamiliar technology but one nearing commercial application. Essentially high-tensile-strength and virtually zero-friction motor-generators, they come in sizes small enough to supply power for cars or houses and also big enough for use by power companies. They will be particularly useful in regions like the Plains, with thin and dispersed populations. Meanwhile, wind is a natural complement to the new natural gas-turbine generators, which turn on and off quickly and can thus make up for drops in wind output.

The costs of building and installing wind turbines have dropped steadily and will certainly continue to drop. Some observers feel that Belgium and Germany have now surpassed the United States in wind technology, and the Japanese are also actively in the race. The authors of a recent study of American competitiveness in environmental industries: "As is the case with other renewable technologies, wind power's early significant advances in this country have led to a worldwide technology development effort that far surpasses current domestic expenditures.

Fig. Modern Wind-energy

Through advanced turbine design, wind power currently costs, on the average, even with coal plants or natural-gas turbines and markedly less expensive than nuclear power.A recent round of competitive bidding for power generation in California showed that wind turbines are already more economical than gas-plant repowering. They are also built very quickly, so when wind power really gets going on the Plains, it is likely to develop with surprising speed.

To assess wind power fully from a sustainability standpoint, a net-energy analysis should be carried out according to the same logic used to assess petroleum-based agriculture. How much energy goes into a wind farm's construction and, later, maintenance, versus how much comes out? A careful study along these lines has been done of photovoltaic cells — the devices now used to generate energy for roadside emergency phones, lighted buoys, and many other remote applications.

Thin-film photovoltaic modules pay off in six months' to two years' time. It is almost certain that wind turbines pay off their energy investment in the first year of operation; thereafter, for the twenty or thirty years that an individual wind machine lasts, it generates net positive energy. Its maintenance causes hardly any environmental degradation, and it requires no mining or drilling, transportation of fuels, emission controls, or waste disposal.

Thus, wind power, as is recognized by most continental European countries and by the British, must be a substantial part of any long-term, sustainable national energy policy. It certainly should come to play a major role on the Plains, as it will in much of the world.

Over the next decade or so, the only way to acquire energy substantially more cheaply than through wind power will be to invest in installing more efficient motors, lighting, heating, and air-conditioning, thus creating "negawatts" — newly available power achieved without the building of new generating plants.

Such conservation investments can make saved power available at a cost some 24-44 percent lower even than that of wind power — something that should have strong appeal to thrifty Plains people. U.S. energy use per capita is so high that conservation is now and will be for some decades the best energy investment we could make, and some state public utility commissions have found inventive ways to motivate utilities to help their customers improve efficiency. Many state regulatory agencies have yet to act on these long-term benefits for their economies, however, so the national pace of change has been regrettably slow. Yet in Iowa, a switch to renewable energy has been adopted as an official state goal, and six wind-power projects are planned or under construction. If the saying "As Iowa goes, so goes the nation" still has validity, this is good news.

WIND POWER ON THE PLAINS

For the Plains, wind has a great many specific advantages over fossil-fuel power. It blows most strongly in the daytime and in the winter, when power needs are highest; it would thus be especially desirable for customers — businesses, farms, or householders — that presently pay stiff rates for peak-period power. Wind installations are highly compatible with ranching operations, whether of cattle or bison.

From a jobs standpoint, residents of the Plains, which have suffered boom-and-bust employment in oil and coal, should find particularly appealing the fact that wind development creates about fifteen jobs for every million dollars of investment, slightly more than coal does,while hydroelectric power and natural gas are only half as job productive.

Plains resources of biomass could also furnish massive amounts of energy through the burning of crop wastes and the cultivation of fast-growing fuel crops like switchgrass and coppiced poplar. Putting biomass and wind together with an aggressive energy-efficiency programme could make the region self-sufficient in energy and result in long-term net savings to the regional economy.

SUSTAINABLE ENERGY FROM MANY SOURCES

Other renewable-energy resources exist in the Plains states and will be developed in time, although wind will lead the way. Hydropower, at dams on the Missouri and other rivers, is already fully exploited and provides the region's cheapest existing electricity. In some areas, like Wyoming, with its cold but relatively cloud-free climate, solar-thermal and photovoltaic possibilities will open up, as well as direct solar power for space and water heating.

In areas of the Plains with substantial rainfall, biomass burning, using either crop wastes or specially grown crops, could yield substantial power without a net addition to atmospheric carbon dioxide, and ethanol production for fuel use is already a major enterprise in Iowa. Wind, however, is the only renewable energy source that leaves the land surface free for grazing — a synergistic relationship that should greatly appeal to landowners.

Deregulation is threatening the previously guaranteed profits of utilities; expectations are clouded. And generating technologies are changing. Despite billions of dollars expended annually in federal subsidies, nuclear power is effectively dead in the United States; it is simply not competitive, even aside from the unsolved — and enormously costly — problem of what to do with its radioactive wastes.

Plants that burn natural gas will remain attractive for a decade or so, but coal plants now carry such heavy costs in pollution cleanup that they too are becoming financially unattractive. Thus, by economic default, as well as by the attractiveness of their pollution-free, durable, no-fuels technologies, solar and wind power have a solid future as parts of America's energy supply system, and the Plains in particular are in a fine position to exploit wind power.

Biomass energy in the form of ethanol already helps to power our vehicles, and we can envision a sustainable Plains world of humans deriving nourishment and economic support from grassland bison and driving vehicles powered by prairie biomass fuels.

Wind also may come to power cars indirectly. Because automobiles and trucks are our greatest users of energy, transportation alternatives to imported oil are a fundamental longterm concern, both nationally and regionally, and wind-generated power will be one component in a future transportation system utilizing mixed energy sources. Our present transportation system gets more than 97 percent of its energy from petroleum — a dangerous dependency.

In the long run, electric vehicles have a bright future because they are much simpler mechanically than internal combustion vehicles, are capable of astonishing acceleration, and have a lower per-mile energy cost, especially when regenerative braking systems capture braking energy and put it back into the battery or flywheel. Electricity for cars can come as easily from wind as from fossil-fuel plants — with great savings in air pollution over internal combustion engines — though it can also be generated by compact internal combustion on-board engines.

Unfortunately, the recently developed and otherwise promising nickel-metal-hydride batteries use toxic nickel, while lead-acid batteries involve some atmospheric pollution in their production and recycling, so flywheel energy storage will probably become very attractive for vehicles. In the distant future, hydrogen will be used for cars and many other purposes since it is a pollution-free fuel and can be piped and stored rather like natural gas. Wind and solar power will ultimately be used to produce hydrogen by dissociating the hydrogen and oxygen that make up water.

Battery-driven electric cars and vans are now becoming common for delivery and other short-haul use. California has required major car companies to sell substantial numbers of electric vehicles in the Los Angeles air-pollution basin by 1997, and pollution-plagued eastern states are adopting similar measures. If electric vehicles prove cheap and reliable, Plains dwellers will begin using them too.

Nevertheless, fossil fuels will undoubtedly continue to provide a good proportion of our transportation energy, even if the government becomes less determined to subsidize the car-highway-oil complex through road building, police and court services, military expenditures to ensure control of the Middle East, and so on.

LIVING WITH WIND POWER

Nothing and nobody is entirely innocent. Blades of wind machines take some toll on hawks, owls, and eagles who try to fly through them, though highways cause roadkills of immensely greater numbers of animals and birds.

To combat bird losses, designers of wind machines are experimenting with slower-moving blades or a more visible and avoidable set of fixed vertical vanes in a star-shaped pattern around a vertical rotor — a system of this type is now in place on a property in Wales belonging to the Queen of England. Wind machines pose no known problems for other animals, particularly large grazers similar to bison — indeed, much experience in California has demonstrated that livestock and wind-power installations coexist very comfortably.

Wind machines make noise — a persistent swishing sound that most people would not enjoy living right next to; it is similar to the wind and tire noise generated by a highway. Luckily, however, the strong winds on hills that attract wind-farm designers also discourage residential siting. On the Plains as elsewhere, few people like to live on hillsides exposed to the strongest winds, so noise exposure should not be a significant problem.

And bison would probably love wind-machine towers for one special bison reason. In the spring, when they are shedding their heavy winter coats, bison like to scratch on vertical objects. On the Plains, this originally meant occasional trees or stumps or rare big rocks, which were polished smooth over centuries and were called "buffalo rocks" by the pioneers. Wind farms would be bison rubbing paradise.

On ranches, bison rubbing can do substantial damage to fence posts. But the bison on Catalina Island have not been a nuisance to the wooden-pole power line crossing the island; nor has the herd at the National Bison Range in Montana damaged the steel power-line towers crossing its territory. So it seems very unlikely that even the most enthusiastic bison rubbing would affect heavy steel wind-machine towers with concrete footings, which are built to withstand severe winds.

Fig. Kenetech Corp.

To me, the sight of a field of wind machines dancing their intricate differential rhythms with the wind is a joy; it always makes me smile. The grave vertical-axis turbines, their great, curved blades revolving slowly like some space-alien eggbeater, fill me with awe.

The whishes and swooshes of wind turbines as the wind direction varies sound reassuring to me, and windy grasslands with thousands of bison happily grazing among wind machines seem to me a delightful prospect. Nonetheless, there are a few people who find wind machines ugly or intrusive on the landscape.

It is hard, of course, to argue about impressions of beauty; as has often been said, beauty lies in the eye of the beholder. If displeased viewers of wind farms go out to observe a remote wind-farm area on foot or bicycle, and at home use only wood power for their heating and cooking and tallow candles for light, their aesthetic criticisms would rest on a purer footing. But it is hard to take such critics seriously when they approach by car over six-lane highways that are far more visible infringements on the landscape than a wind farm — besides being substantial contributors to atmospheric pollution.

Moreover, unless the critics are very unusual indeed, they are (like all of us) heavy users of natural gas, oil, and electricity generated by burning of fossil fuels or deployment of radioactivity. A visit to a refinery, coal mine, or nuclear plant should be a prerequisite before disparaging a relatively benign, life compatible, and sustainable technology like wind power.

Chapter 9

Nuclear Energy-Fission and Fusion

Nuclear energy is the energy that is trapped inside each atom. The ancient Greeks believed that the smallest part of nature is an atom. But they did not know 2000 years ago that atoms are made up of further smaller particles—a nucleus of protons and neutrons, surrounded by electrons, which swirl around the nucleus much like the earth revolves around the sun. One of the laws of the universe is that matter and energy can neither be created nor destroyed. But they can be changed in form. Matter can be changed into energy.

Albert Einstein's famous mathematical formula $E = mc^2$ explains this. The equation says: E [energy] equals m [mass] times c^2 [c stands for the speed or velocity of light]. This means that it is mass multiplied by the square of the velocity of light.Scientists used Einstein's equation as the key to unlock atomic energy and to create atomic bombs.An atom's nucleus can be split apart.

This is known as fission. When this is done, a tremendous amount of energy in the form of both heat and light is released by the initiation of a chain reaction. This energy, when slowly released, can be harnessed to generate electricity. When it is released all at once, it results in a tremendous explosion as in an atomic bomb.

Nuclear energy can also be harnessed by fusion. A fusion reaction occurs when two hydrogen atoms combine to produce one helium atom. This reaction takes place at all times in the sun, which provides us with the solar energy. This technology is still at the experimental stage and may become viable in future. Uranium is the main element required to run a nuclear reactor where energy is extracted. Uranium is mined from many places around the world. It is processed (to get enriched uranium, i.e. the radioactive isotope) into tiny pellets. These pellets are loaded into long rods that are put into the power plant's reactor.

Inside the reactor of an atomic power plant, uranium atoms are split apart in controlled chain reaction. Other fissile material includes plutonium and thorium.In a chain reaction, particles released by the splitting of the atom strike other uranium atoms and split them. The particles released by this further split other atoms in a chain process. In nuclear power plants, control rods are used to keep the splitting regulated, so that it does not occur too fast. These are called moderators.

The chain reaction gives off heat energy. This heat energy is used to boil heavy water in the core of the reactor. So, instead of burning a fuel, nuclear power plants use the energy released by the chain reaction to change the energy of atoms into heat energy. The heavy water from around the nuclear core is sent to another section of the power plant.

Here it heats another set of pipes filled with water to make steam. The steam in this second set of pipes rotates a turbine to generate electricity. If the reaction is not controlled, you could have an atomic bomb.But in atomic bombs, almost pure pieces of uranium-235 or plutonium, of a precise mass and shape, must be brought together and held together with great force. These conditions are not present in a nuclear reactor.The reaction also creates radioactive material.

This material could hurt people if released, so it is kept in a solid form. A strong concrete dome is built around the reactor to prevent this material from escaping in case of an accident.Experiences with nuclear programmes differ and the future of nuclear power remains uncertain because of public reaction. But in the past few years the capacity of operating nuclear plants has increased more than twentyfold. There are more than 400 nuclear power plants providing about 7% of the world's primary energy and about 25% of the electric power in industrialized nations.

The growth of nuclear power combined with the shift

from carbon-heavy fuels such as coal and oil to carbon-light gas contribute to the gradual 'de-carbonization' of the world energy system.

Chernobyl, Three Mile Island, and other nuclear accidents have increased the fear of harnessing nuclear energy. Another issue with international and local implications is the storage and disposal of radioactive wastes: both from nuclear reactors making electricity and from the production of military weapons.

Earlier disposal practices, such as dumping of nuclear waste at sea, have been completely stopped by formal treaty because of environmental concerns (and by cessation of furtive scuttling of nuclear submarines). Regimes for transport and temporary storage of civil and defence nuclear wastes now function, although sites and designs for permanent disposal have yet to be accepted.

Nuclear Energy is released by the splitting (fission) or merging together (fusion) of the nuclei of atom(s). The conversion of nuclear mass to energy is consistent with the mass-energy equivalence formula $\Delta E = \Delta m.c^2$, in which ΔE = energy release, Δm = mass defect, and c = the speed of light in a vacuum (a physical constant).

Nuclear energy was first discovered by French physicist Henri Becquerel in 1896, when he found that photographic plates stored in the dark near uranium were blackened like X-ray plates, which had been just recently discovered at the time 1895. Nuclear chemistry can be used as a form of alchemy to turn lead into gold or change any atom to any other atom (albeit through many steps). Radionuclide (radioisotope) production often involves irradiation of another isotope (or more precisely a nuclide), with alpha particles, beta particles, or gamma rays.

Iron has the highest binding energy per nucleon of any atom. If an atom of lower average binding energy is changed into an atom of higher average binding energy, energy is given off. The chart shows that fusion of hydrogen, the combination to form heavier atoms, releases energy, as does fission of uranium, the breaking up of a larger nucleus into smaller parts.

Stability varies between isotopes: the isotope U-235 is much less stable than the more common U-238.

THE ATOM

The atom consists of a small, massive, positively charged core (nucleus) surrounded by electrons. The nucleus, containing most of the mass of the atom, is itself composed of neutrons and protons bound together by very strong nuclear forces, much greater than the electrical forces that bind the electrons to the nucleus.

The mass number A of a nucleus is the number of nucleons, or protons and neutrons, it contains; the atomic number Z is the number of positively charged protons. A specific nucleus is designated as ¿U.

The expression ¯U, represents uranium-235. The binding energy of a nucleus is a measure of how tightly its protons and neutrons are held together by the nuclear forces. The binding energy per nucleon, the energy required to remove one neutron or proton from a nucleus, is a function of the mass number A.

The curve of binding energy implies that if two light nuclei near the left end of the curve coalesce to form a heavier nucleus, or if a heavy nucleus at the far right splits into two lighter ones, more tightly bound nuclei result, and energy will be released.Nuclear energy, measured in millions of electron volts (MeV), is released by the fusion of two light nuclei, as when two heavy hydrogen nuclei, deuterons (ªH), combine in the reaction

$$^{2}_{1}\mathrm{H} + ^{2}_{1}\mathrm{H} \rightarrow ^{3}_{2}\mathrm{He} + ^{1}_{0}n + 3.2\mathrm{MeV}$$

producing a helium-3 atom, a free neutron (¦n), and 3.2 MeV, or 5.1×10^{-13} J (1.2×10^{-13} cal). Nuclear energy is also released when the fission of a heavy nucleus such as ¯U is induced by the absorption of a neutron as in

$$^{1}_{0}\mathrm{n} + ^{235}_{92}\mathrm{U} \rightarrow ^{140}_{55}\mathrm{Cs} + ^{93}_{37}\mathrm{Rb} + 3^{1}_{0}\mathrm{n} + 200\mathrm{MeV}$$

producing cesium-140, rubidium-93, three neutrons, and 200 MeV, or 3.2×10^{-11} J (7.7×10^{-12} cal). A nuclear fission reaction releases 10 million times as much energy as is released in a typical chemical reaction.

Production

Changes can occur in the structure of the nuclei of atoms. These changes are called nuclear reactions. Energy created in a nuclear reaction is called nuclear energy, or atomic energy.

Nuclear energy is produced naturally and in man-made operations under human control.

- *Naturally:* Some nuclear energy is produced naturally.The Sun and other stars make heat and light by nuclear reactions.
- *Man-Made:* Nuclear energy can be man-made too. Machines called nuclear reactors, parts of nuclear power plants, provide electricity for many cities. Man-made nuclear reactions also occur in the explosion of atomic and hydrogen bombs.

Nuclear energy is produced in two different ways, in one, large nuclei are split to release energy. In the other method, small nuclei are combined to release energy.

- *Nuclear Fission:* In nuclear fission, the nuclei of atoms are split, causing energy to be released. The atomic bomb and nuclear reactors work by fission. The element uranium is the main fuel used to undergo nuclear fission to produce energy since it has many favorable properties. Uranium nuclei can be easily split by shooting neutrons at them. Also, once a uranium nucleus is split, multiple neutrons are released which are used to split other uranium nuclei. This phenomenon is known as a chain reaction.

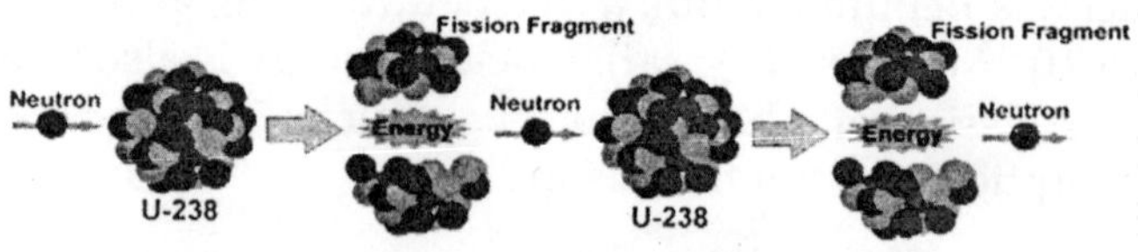

Fig. Fission of Uranium 235 Nucleus.

- *Nuclear Fusion:* In nuclear fusion, the nuclei of atoms are joined together, or fused. This happens only under very hot conditions. The Sun, like all other stars, creates heat and light through nuclear fusion. In the Sun, hydrogen nuclei fuse to make helium. The hydrogen bomb, humanity's most powerful and destructive weapon, also works by fusion. The heat required to start the fusion reaction is so great that an atomic bomb is used to provide it. Hydrogen nuclei fuse to form helium and in the process release huge amounts of energy thus producing a huge explosion.

Advantages of Nuclear Energy

- The Earth has limited supplies of coal and oil. Nuclear power plants could still produce electricity after coal and oil become scarce.
- Nuclear power plants need less fuel than ones which burn fossil fuels. One ton of uranium produces more energy than is produced by several million tons of coal or several million barrels of oil.
- Coal and oil burning plants pollute the air. Well-operated nuclear power plants do not release contaminants into the environment.

Disadvantages of Nuclear Energy

The nations of the world now have more than enough nuclear bombs to kill every person on Earth. The two most powerful nations – Russia and the United States – have about 50,000 nuclear weapons between them. Nuclear explosions produce radiation. The nuclear radiation harms the cells of the body which can make people sick or even kill them. Illness can strike people years after their exposure to nuclear radiation.

- One possible type of reactor disaster is known as a meltdown. In such an accident, the fission reaction goes out of control, leading to a nuclear explosion and the emission of great amounts of radiation.
- The cooling system failed at the Three Mile Island nuclear reactor near Harrisburg, Pennsylvania. Radiation leaked, forcing tens of thousands of people

to flee. The problem was solved minutes before a total meltdown would have occurred. Fortunately, there were no deaths.

- Nuclear reactors also have waste disposal problems. Reactors produce nuclear waste products which emit dangerous radiation. Because they could kill people who touch them, they cannot be thrown away like ordinary garbage. Currently, many nuclear wastes are stored in special cooling pools at the nuclear reactors.
- Nuclear reactors only last for about forty to fifty years.

NUCLEAR ENERGY FROM FISSION

The two key characteristics of nuclear fission important for the practical release of nuclear energy are both evident. First, the energy per fission is very large. In practical units, the fission of 1 kg (2.2 lb) of uranium-235 releases 18.7 million kilowatt-hours as heat. Second, the fission process initiated by the absorption of one neutron in uranium-235 releases about 2.5 neutrons, on the average, from the split nuclei. The neutrons released in this manner quickly cause the fission of two more atoms, thereby releasing four or more additional neutrons and initiating a self-sustaining series of nuclear fissions, or a chain reaction, which results in continuous release of nuclear energy.

Naturally occurring uranium contains only 0.71 percent uranium-235; the remainder is the nonfissile isotope uranium-238. A mass of natural uranium by itself, no matter how large, cannot sustain a chain reaction because only the uranium-235 is easily fissionable. The probability that a fission neutron with an initial energy of about 1 MeV will induce fission is rather low, but the probability can be increased by a factor of hundreds when the neutron is slowed down through a series of elastic collisions with light nuclei such as hydrogen, deuterium, or carbon. This fact is the basis for the design of practical energy-producing fission reactors.

Nuclear Power Reactors

The first large-scale nuclear reactors were built in 1944 at

Hanford, Washington, for the production of nuclear weapons material. The fuel was natural uranium metal; the moderator, graphite. Plutonium was produced in these plants by neutron absorption in uranium-238; the power produced was not used.

LIGHT-WATER AND HEAVY-WATER REACTORS

A variety of reactor types, characterized by the type of fuel, moderator, and coolant used, have been built throughout the world for the production of electric power. In the United States, with few exceptions, power reactors use nuclear fuel in the form of uranium oxide isotopically enriched to about three percent uranium-235. The moderator and coolant are highly purified ordinary water. A reactor of this type is called a light-water reactor (LWR).

In the pressurized-water reactor (PWR), a version of the LWR system, the water coolant operates at a pressure of about 150 atmospheres. It is pumped through the reactor core, where it is heated to about 325° C (about 620° F). The superheated water is pumped through a steam generator, where, through heat exchangers, a secondary loop of water is heated and converted to steam.

This steam drives one or more turbine generators, is condensed, and is pumped back to the steam generator. The secondary loop is isolated from the water in the reactor core and, therefore, is not radioactive. A third stream of water from a lake, river, or cooling tower is used to condense the steam. The reactor pressure vessel is about 15 m (about 49 ft) high and 5 m (about 16.4 ft) in diameter, with walls 25 cm (about 10 in) thick. The core houses some 82 metric tons of uranium oxide contained in thin corrosion-resistant tubes clustered into fuel bundles.

In the boiling-water reactor (BWR), a second type of LWR, the water coolant is permitted to boil within the core, by operating at somewhat lower pressure. The steam produced in the reactor pressure vessel is piped directly to the turbine generator, is condensed, and is then pumped back to the reactor. Although the steam is radioactive, there is no intermediate heat exchanger between the reactor and turbine to decrease efficiency. As in the PWR, the condenser cooling water has a separate source, such as a lake or river.

The power level of an operating reactor is monitored by a variety of thermal, flow, and nuclear instruments. Power output is controlled by inserting or removing from the core a group of neutron-absorbing control rods. The position of these rods determines the power level at which the chain reaction is just self-sustaining.During operation, and even after shutdown, a large, 1,000-megawatt (MW) power reactor contains billions of curies of radioactivity. Radiation emitted from the reactor during operation and from the fission products after shutdown is absorbed in thick concrete shields around the reactor and primary coolant system. Other safety features include emergency core cooling systems to prevent core overheating in the event of malfunction of the main coolant systems and, in most countries, a large steel and concrete containment building to retain any radioactive elements that might escape in the event of a leak.

Although more than 100 nuclear power plants were operating or being built in the United States at the beginning of the 1980s, in the aftermath of the Three Mile Island accident in Pennsylvania in 1979 safety concerns and economic factors combined to block any additional growth in nuclear power.

No orders for nuclear plants have been placed in the United States since 1978, and some plants that have been completed have not been allowed to operate. In 1996 about 22 percent of the electric power generated in the United States came from nuclear power plants. In contrast, in France almost three-quarters of the electricity generated was from nuclear power plants.

Propulsion Reactors

Nuclear power plants similar to the PWR are used for the propulsion plants of large surface naval vessels such as the aircraft carrier USS *Nimitz.* The basic technology of the PWR system was first developed in the U.S. naval reactor programme directed by Admiral Hyman G. Rickover. Reactors for submarine propulsion are generally physically smaller and use more highly enriched uranium to permit a compact core. The United States, the United Kingdom, Russia, and France all have nuclear-powered submarines with such power plants.

Three experimental seagoing nuclear cargo ships were operated for limited periods by the United States, Germany, and Japan. Although they were technically successful, economic conditions and restrictive port regulations brought an end to these projects. The Soviet government built the first successful nuclear-powered icebreaker, *Lenin,* for use in clearing the Arctic sea-lanes.

Research Reactors

A variety of small nuclear reactors have been built in many countries for use in education and training, research, and the production of radioactive isotopes. These reactors generally operate at power levels near one MW, and they are more easily started up and shut down than larger power reactors.

A widely used type is called the swimming-pool reactor. The core is partially or fully enriched uranium-235 contained in aluminum alloy plates, immersed in a large pool of water that serves as both coolant and moderator. Materials may be placed directly in or near the reactor core to be irradiated with neutrons. Various radioactive isotopes can be produced for use in medicine, research, and industry. Neutrons may also be extracted from the reactor core by means of beam tubes to be used for experimentation.

BREEDER REACTORS

Uranium, the natural resource on which nuclear power is based, occurs in scattered deposits throughout the world. Its total supply is not fully known, and may be limited unless sources of very low concentration such as granites and shale were to be used. Conservatively estimated U.S. resources of uranium having an acceptable cost lie in the range of two million to five million metric tons.

The lower amount could support an LWR nuclear power system providing about 30 percent of U.S. electric power for only about 50 years. The principal reason for this relatively brief life span of the LWR nuclear power system is its very low efficiency in the use of uranium: only approximately one percent of the energy content of the uranium is made available in this system.

The key feature of a breeder reactor is that it produces more fuel than it consumes. It does this by promoting the absorption of excess neutrons in a fertile material. Several breeder reactor systems are technically feasible. The breeder system that has received the greatest worldwide attention uses uranium-238 as the fertile material. When uranium-238 absorbs neutrons in the reactor, it is transmuted to a new fissionable material, plutonium, through a nuclear process called *â* (beta) decay. The sequence of nuclear reactions is

$$^{235}_{92}U + ^{1}_{0}n \longrightarrow ^{239}_{92}U \xrightarrow[\beta]{} ^{239}_{92}Np \xrightarrow[\beta]{} ^{239}_{94}Nu$$

In beta decay a nuclear neutron decays into a proton and a beta particle (a high-energy electron). When plutonium-239 itself absorbs a neutron, fission can occur, and on the average about 2.8 neutrons are released. In an operating reactor, one of these neutrons is needed to cause the next fission and keep the chain reaction going. On the average about 0.5 neutron is uselessly lost by absorption in the reactor structure or coolant.

The remaining 1.3 neutrons can be absorbed in uranium-238 to produce more plutonium via the reactions in equation (3).The breeder system that has had the greatest development effort is called the liquid-metal fast breeder reactor (LMFBR). In order to maximize the production of plutonium-239, the velocity of the neutrons causing fission must remain fast—at or near their initial release energy. Any moderating materials, such as water, that might slow the neutrons must be excluded from the reactor.

A molten metal, liquid sodium, is the preferred coolant liquid. Sodium has very good heat transfer properties, melts at about 100° C (about 212° F), and does not boil until about 900° C (about 1650° F). Its main drawbacks are its chemical reactivity with air and water and the high level of radioactivity induced in it in the reactor.Development of the LMFBR system began in the United States before 1950, with the construction of the first experimental breeder reactor, EBR-1. A larger U.S. programme, on the Clinch River, was halted in 1983, and only experimental work was to continue. In one design of a large LMFBR power plant, the core of the reactor consists of thousands of thin stainless steel tubes containing mixed uranium and plutonium oxide fuel: about 15 to 20 percent plutonium-239, the remainder uranium.

Surrounding the core is a region called the breeder blanket, which contains similar rods filled only with uranium oxide. The entire core and blanket assembly measures about 3 m (about 10 ft) high by about 5 m (about 16.4 ft) in diameter and is supported in a large vessel containing molten sodium that leaves the reactor at about 500° C (about 930° F). This vessel also contains the pumps and heat exchangers that aid in removing heat from the core. Steam is produced in a second sodium loop, separated from the radioactive reactor coolant loop by the intermediate heat exchangers in the reactor vessel. The entire nuclear reactor system is housed in a large steel and concrete containment building.

The first large-scale plant of this type for the generation of electricity, called Super-Phénix, went into operation in France in 1984. (However, concerns about operational safety and environmental contamination led the French government to announce in 1998 that Super-Phénix would be dismantled). An intermediate-scale plant, the BN-600, was built on the shore of the Caspian Sea for the production of power and the desalination of water. The British have a large 250-MW prototype in Scotland.

The LMFBR produces about 20 percent more fuel than it consumes. In a large power reactor enough excess new fuel is produced over 20 years to permit the loading of another similar reactor. In the LMFBR system about 75 percent of the energy content of natural uranium is made available, in contrast to the one percent in the LWR.

Nuclear Fuels and Wastes

The hazardous fuels used in nuclear reactors present

handling problems in their use. This is particularly true of the spent fuels, which must be stored or disposed of in some way.

The Nuclear Fuel Cycle

Any electric power generating plant is only one part of a total energy cycle. The uranium fuel cycle that is employed for LWR systems currently dominates worldwide nuclear power production and includes many steps. Uranium, which contains about 0.7 percent uranium-235, is obtained from either surface or underground mines.

The ore is concentrated by milling and then shipped to a conversion plant, where its elemental form is changed to uranium hexafluoride gas (UF_6). At an isotope enrichment plant, the gas is forced against a porous barrier that permits the lighter uranium-235 to penetrate more readily than uranium-238. This process enriches uranium to about 3 percent uranium-235.

The depleted uranium—the tailings—contain about 0.3 percent uranium-235. The enriched product is sent to a fuel fabrication plant, where the UF_6 gas is converted to uranium oxide powder, then into ceramic pellets that are loaded into corrosion-resistant fuel rods. These are assembled into fuel elements and are shipped to the reactor power plant. A typical 1,000-MW pressurized-water reactor has about 200 fuel elements, one-third of which are replaced each year because of the depletion of the uranium-235 and the buildup of fission products that absorb neutrons. At the end of its life in the reactor, the fuel is tremendously radioactive because of the fission products it contains and hence is still producing a considerable amount of energy. The discharged fuel is placed in water storage pools at the reactor site for a year or more.

At the end of the cooling period the spent fuel elements are shipped in heavily shielded casks either to permanent storage facilities or to a chemical reprocessing plant. At a reprocessing plant, the unused uranium and the plutonium-239 produced in the reactor are recovered and the radioactive wastes concentrated.

The spent fuel still contains almost all the original

uranium-238, about one-third of the uranium-235, and some of the plutonium-239 produced in the reactor. In cases where the spent fuel is sent to permanent storage, none of this potential energy content is used. In cases where the fuel is reprocessed, the uranium is recycled through the diffusion plant, and the recovered plutonium-239 may be used in place of some uranium-235 in new fuel elements. At the end of the 20th century, no reprocessing of fuel occurred in the United States because of environmental, health, and safety concerns, and the concern that plutonium-239 could be used illegally for the manufacture of weapons.

In the fuel cycle for the LMFBR, plutonium bred in the reactor is always recycled for use in new fuel. The feed to the fuel-element fabrication plant consists of recycled uranium-238, depleted uranium from the isotope separation plant stockpile, and part of the recovered plutonium-239. No additional uranium needs to be mined, as the existing stockpile could support many breeder reactors for centuries. Because the breeder produces more plutonium-239 than it requires for its own refueling, about 20 percent of the recovered plutonium is stored for later use in starting up new breeders. Because new fuel is bred from the uranium-238, instead of using only the natural uranium-235 content, about 75 percent of the potential energy of uranium is made available with the breeder cycle.

The final step in any of the fuel cycles is the long-term storage of the highly radioactive wastes, which remain biologically hazardous for thousands of years.

Fuel elements may be stored in shielded, guarded repositories for later disposition or may be converted to very stable compounds, fixed in ceramics or glass, encapsulated in stainless steel canisters, and buried far underground in very stable geologic formations. However, the safety of such repositories is the subject of public controversy, especially in the geographic region in which the repository is located or is proposed to be built. Environmentalists plan to file a lawsuit to close a repository built near Carlsbad, New Mexico.

In 1999, this repository began receiving shipments of radioactive waste from the manufacture of nuclear weapons in United States during the Cold War. Another controversy centers around a proposed repository at Yucca Mountain, Nevada. Opposition from state residents and questions about the geologic stability of this site have helped prolong government studies. Even if opened, the site will not receive shipments of radioactive waste until at least 2010.

FUEL REPROCESSING

The fuel reprocessing step poses a combination of radiological hazards. One is the accidental release of fission products if a leak should occur in chemical equipment or the cells and building housing it. Another may be the routine release of low levels of inert radioactive gases such as xenon and krypton. In 1966 a commercial reprocessing plant opened in West Valley, New York. But in 1972 this reprocessing plant was closed after generating more than 600,000 gallons of high-level radioactive waste.

After the plant was closed, a portion of this radioactive waste was partially treated and cemented into nearly 20,000 steel drums. In 1996, the United States Department of Energy began to solidify the remaining liquid radioactive wastes into glass cylinders. At the end of the 20th century, no reprocessing plants were licensed in the United States.

Of major concern in chemical reprocessing is the separation of plutonium-239, a material that can be used to make nuclear weapons. The hazards of theft of plutonium-239, or its use for intentional but hidden production for weapons purposes, can best be controlled by political rather than technical means. Improved security measures at sensitive points in the fuel cycle and expanded international inspection by the International Atomic Energy Agency (IAEA) offer the best prospects for controlling the hazards of plutonium diversion.

WASTE MANAGEMENT

The last step in the nuclear fuel cycle, waste management, remains one of the most controversial. The principal issue here is not so much the present danger as the danger to generations far in the future. Many nuclear wastes remain radioactive for thousands of years, beyond the span of any human institution. The technology for packaging the wastes so that they pose no current hazard is relatively straightforward. The difficulty lies both in being adequately confident that future generations are well protected and in making the political decision on how and where to proceed with waste storage. Permanent but potentially retrievable storage in deep stable geologic formations seems the best solution.

In 1988 the U.S. government chose Yucca Mountain, a Nevada desert site with a thick section of porous volcanic rocks, as the nation's first permanent underground repository for more than 36,290 metric tons of nuclear waste. However, opposition from state residents and uncertainty that Yucca Mountain may not be completely insulated from earthquakes and other hazards has prolonged government studies.

The presence of water in these samples suggests that water may have once risen up through the mountain and later subsided. Because such an event could jeopardize the safety of a nuclear waste repository, the Department of Energy has funded more study of these fluid intrusions.

URANIUM PREPARATION

Earlier we talked about nuclear fission with 235U. In reality, this will not be the only isotope of uranium present in a nuclear reactor. In naturally occurring uranium deposits, less than one percent of the uranium is 235U. The majority of the uranium is 238U. 238U is not a fissile isotope of uranium. When 238U is struck by a loose neutron, it absorbs the neutron into its nucleus and does not fission.

Thus, by absorbing loose neutrons, 238U can prevent a nuclear chain reaction from occurring. This would be a bad thing because if a chain reaction doesn't occur, the nuclear reactions can't sustain themselves, the reactor shuts down, and millions of people are without electrical power.

In order for a chain reaction to occur, the pure uranium ore must be refined to raise the concentration of 235U. This is called enrichment and is primarily accomplished through a technique called gaseous diffusion. In this process, the uranium ore is combined with fluorine to create a chemical compound called uranium hexafluoride. The uranium hexafluoride is heated and vaporizes.

The heated gas is then pushed through a series of filters. Because some of the uranium hexafluoride contains 238U and some contains 235U, there is a slight difference in the weights of the individual molecules. The molecules of uranium hexafluoride containing 235U are slightly lighter and thus pass more easily through the filters. This creates a quantity of uranium hexafluoride with a higher proportion of 235U. This is collected, the uranium is stripped from it, and the result is an enriched supply of fuel. Usually, nuclear power plants use uranium fuel that is about 4% 235U.

PARTS OF A NUCLEAR REACTOR

A typical nuclear reactor has a few main parts. Inside the "core" where the nuclear reactions take place are the fuel rods and assemblies, the control rods, the moderator, and the coolant. Outside the core are the turbines, the heat exchanger, and part of the cooling system.

The fuel assemblies are collections of fuel rods. These rods are each about 3.5 meters (11.48 feet) long. They are each about a centimeter in diameter. These are grouped into large bundles of a couple hundred rods called fuel assemblies, which are then placed in the reactor core. Inside each fuel rod are hundreds of pellets of uranium fuel stacked end to end.

Also in the core are control rods. These rods have pellets inside that are made of very efficient neutron capturers. An example of such a material is cadmium. These control rods are connected to machines that can raise or lower them in the core. When they are fully lowered into the core, fission can not occur because they absorb free neutrons. However, when they are pulled out of the reactor, fission can start again anytime a stray neutron strikes a 235U atom, thus releasing more neutrons, and starting a chain reaction.

Another component of the reactor is the moderator. The moderator serves to slow down the high speed neutrons "flying" all around the reactor core. If a neutron is moving too fast, and thus is at a high-energy state, it passes right through the 235U nucleus. It must be slowed down to be captured by the nucleus and to induce fission. The most common moderator is water, but sometimes it can be another material.

The job of the coolant is to absorb the heat from the reaction. The most common coolant used in nuclear power plants today is water. In actuality, in many reactor designs the coolant and the moderator are one and the same. The coolant water is heated by the nuclear reactions going on inside the core. However, this heated water does not boil because it is kept at an extremely intense pressure, thus raising its boiling point above the normal 100° Celsius.

The heated water rises up and passes through another part of the reactor, the heat exchanger. The moderator/coolant water is radioactive, so it can not leave the inner reactor containment. Its heat must be transferred to non-radioactive water, which can then be sent out of the reactor shielding. This is done through the heat exchanger, which works by moving the radioactive water through a series of pipes that are wrapped around other pipes. The metallic pipes conduct the heat from the moderator to the normal water. Then, the normal water (now in steam form and intensely hot) moves to the turbine, where electricity is produced.

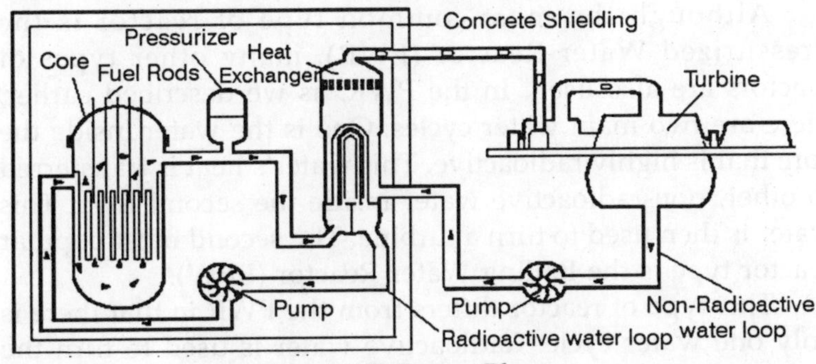

Fig. PWR Power Plant.

After the hot water has passed through the turbine, some of its energy is changed into electricity. However, the water is still very hot. It must be cooled somehow. Many nuclear power plants used steam towers to cool this water with air. These are generally the buildings that people associate with nuclear power plants. At reactors that do not have towers, the clean water is purified and dumped into the nearest body of water, and cool water is pumped in to replace it.

From Fission to Electricity:

A nuclear power plant produces electricity in almost exactly the same way that a conventional (fossil fuel) power plant does. A conventional power plant burns fuel to create heat. The fuel is generally coal, but oil is also sometimes used. The heat is used to raise the temperature of water, thus causing it to boil. The high temperature and intense pressure steam that results from the boiling of the water turns a turbine, which then generates electricity.

A nuclear power plant works the same way, except that the heat used to boil the water is produced by a nuclear fission reaction using 235U as fuel, not the combustion of fossil fuels. A nuclear power plant uses much less fuel than a comparable fossil fuel plant. A rough estimate is that it takes 17,000 kilograms of coal to produce the same amount of electricity as 1 kilogram of nuclear uranium fuel.

Other Types of Reactors:

Although the most common type of reactor is the Pressurized Water Reactor (PWR), many other types of reactors are also used. In the PWR, as we described earlier, there are two main water cycles. One is the water inside the core that is highly radioactive. This water's heat is transferred to other, non-radioactive water inside the second loop. This water is then used to turn a turbine. The second most popular reactor type is the Boiling Water Reactor (BRW).

This type of reactor differs from the PWR in that there is only one water cycle. Radioactive water is used to turn the turbine. The major disadvantage of this is that the radioactive nuclides in the water that cause its radioactivity can be transferred to the turbine, thus causing it to become radioactive too. This produces more hazardous material that needs to be disposed of when a reactor is dismantled. However, the BWR also has a few advantages. Its core can be kept at a lower pressure, for example.

Another type of reactor is the Heavy Water Reactor (HWR). A HWR uses heavy water as a moderator instead of normal water. Heavy water is water with deuterium, which is an isotope of hydrogen with 1 neutron. Deuterium is heavier than normal hydrogen, which has no neutrons. HWR's come in two types, pressurized and boiling, just like normal "light water" reactors.

The advantage of a HWR is that un-enriched uranium fuel can be used. This is because the heavy water is a much more efficient moderator than light water. Thus, more stray neutrons can be slowed down enough to cause fission in 235U. This more efficient moderator makes up for the greater abundance of the neutron-capturing 238U.

Nuclear Fission

If a massive nucleus like uranium-235 breaks apart (fissions), then there will be a net yield of energy because the sum of the masses of the fragments will be less than the mass of the uranium nucleus. If the mass of the fragments is equal to or greater than that of iron at the peak of the binding energy curve, then the nuclear particles will be more tightly bound than they were in the uranium nucleus, and that decrease in mass comes off in the form of energy according to the Einstein equation. For elements lighter than iron, fusion will yield energy. The fission of U-235 in reactors is triggered by the absorption of a low energy neutron, often termed a "slow neutron" or a "thermal neutron". Other fissionable isotopes which can be induced to fission by slow neutrons are plutonium-239, uranium-233, and thorium-232.

Uranium-235 Fission

In one of the most remarkable phenomena in nature, a slow neutron can be captured by a uranium-235 nucleus, rendering it unstable toward nuclear fission. A fast neutron will not be captured, so neutrons must be slowed down by moderation to increase their capture probability in fission reactors. A single fision event can yield over 200 million times the energy of the neutron which triggered it!

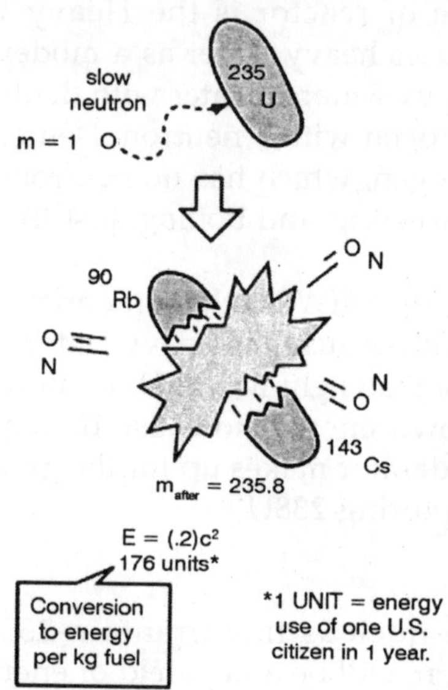

Uranium Fuel

Natural uranium is composed of 0.72% U-235 (the fissionable isotope), 99.27% U-238, and a trace quantity 0.0055% U-234. The 0.72% U-235 is not sufficient to produce a self-sustaining critical chain reaction in U.S. style light-water reactors, although it is used in Canadian CANDU reactors.

For light-water reactors, the fuel must be enriched to 2.5-3.5% U-235. Uranium is found as uranium oxide which when purified has a rich yellow colour and is called "yellowcake". After reduction, the uranium must go through an isotope enrichment process. Even with the necessity of enrichment, it still takes only about 3 kg of natural uranium to supply the energy needs of one American for a year.

Fissionable Isotopes

While uranium-235 is the naturally occuring fissionable

isotope, there are other isotopes which can be induced to fission by neutron bombardment. Plutonium-239 is also fissionable by bombardment with slow neutrons, and both it and uranium-235 have been used to make nuclear fission bombs. Plutonium-239 can be produced by "breeding" it from uranium-238.

Uranium-238, which makes up 99.3% of natural uranium, is not fissionable by slow neutrons. U-238 has a small probability for spontaneous fission and also a small probability of fission when bombarded with fast neutrons, but it is not useful as a nuclear fuel source.

Some of the nuclear reactors at Hanford, Washington and the Savannah-River Plant (SC) are designed for the production of bomb-grade plutonium-239. Thorium-232 is fissionable, so could conceivably be used as a nuclear fuel. The only other isotope which is known to undergo fission upon slow-neutron bombardment is uranium-233.

History of U-235 Fission

German physicists/chemists Otto Hahn and Fritz Strassman attempted to create transuranic elements by bombarding uranium with neutrons. Rather than the heavy elements they expected, they got several unidentified products. When they finally identified one of the products as Barium-141, they were reluctant to publish the finding because it was so unexpected.

When they finally published the results in 1939, they came to the attention of Lise Meitner, an Austrian-born physicist who had worked with Hahn on his nuclear experiments. Upon Hitler's invasion of Austria, she had been forced to flee to Sweden where she and Otto Frisch, her nephew, continued to work on the neutron bombardment problem. She was the first to realise that Hahn's barium and other lighter products from the neutron bombardment experiments were coming from the fission of U-235.

Frisch and Meitner carried out further experiments which showed that the U-235 fission yielded an enormous amount of energy, and that the fission yielded at least two neutrons per neutron absorbed in the interaction. They realized that this made possible a chain reaction with an unprecedented energy yield.

Some Nuclear Units

Nuclear energies are very high compared to atomic processes, and need larger units

1 electron volt = 1 eV = 1.6×10^{-19} joules

1 MeV eV 1 GeV = 10^9 eV 1 TeV = 10^{-12} eV

However, the sizes are quite small and need smaller units:

Atomic size are on the order of 1Å = 10^{-10} m

Nuclear sizes are on the order of femtometers which in the nuclear context are usually called fermis: 1 fm = 10^{-15} m

Nuclear masses are measured in terms of atomic mass units with the carbon-12 nucleus defined as having a mass of exactly 12 amu. It is also common practice to quote the rest mass energy $E=m_0c^2$ as if it were the mass. The conversion to amu is:

$$1u = 1.66054 \times 10^{-27}\ kg = 931.494\ MeV$$

Nuclear Size and Density

Various types of scattering experiments suggest that nuclei are roughly spherical and appear to have essentially the same density. The data are summarized in the expression called the Fermi model:

$$r = r_0 A^{1/3} \text{where } r_0 = 1.2x\ 10^{-15}\ m = 1.2\ fm$$

where r is the radius of the nucleus of mass number A. The assumption of constant density leads to a nuclear density

$$\rho_n = 2.3x\ 10^{17}\ kg/m^3$$

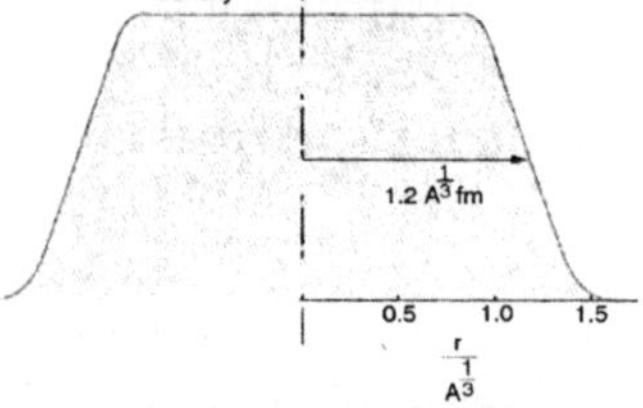

The most definitive information about nuclear sizes comes

from electron scattering. The comparison of calculated and experimental radii for nuclei are very sensitive to the exact onset of the overlap between the probe particle and the nuclear matter. These comparisons have made it clear that there is a "tail" where the density of nuclear matter decreases toward zero. The nucleus is not a hard sphere.

Krane comments that the evidence points to a mass radius and a charge radius which agree with each other within about 0.1 fermi. Since heavy nuclei have about 50% more neutrons than protons, one might expect a mass radius larger than the charge radius. One can visualize the protons being pushed toward the outside by the proton repulsion and the neutrons being pulled inward by the neutron-proton attraction, so the observed result agrees with what one might expect with this kind of model.

Nuclear Density and the Strong Force

The fact that the nuclear density seems to be independent of the details of neutron number or proton number implies that the force between the particles is essentially the same whether they are protons or neutrons. This correlates with other evidence that the strong force is the same between any pair of nucleons.

Nuclear Particles

Nuclei are made up of protons and neutrons bound together by the strong force. Both protons and neutrons are referred to as nucleons. The number of protons is called the atomic number and determines the chemical element. Nuclei of a given element (same atomic number) may have different numbers of neutrons and are then said to be different isotopes of the element.

Electron Scattering from Nuclei

The scattering of electrons from nuclei has given us the most precise information about nuclear size and charge distribution. The electron is a better nuclear probe than the alpha particles of Rutherford scattering because it is a point particle and can penetrate the nucleus. For low energies and under conditions where the electron does not penetrate the nucleus, the electron scattering can be described by the Rutherford formula.

The Rutherford formula is an analytic expression for the differential scattering cross section, and for a projectile charge of 1, it is

$$\left(\frac{d\sigma}{d\cos\theta}\right)_R = \frac{\pi}{2} Z^2 \alpha^2 \left(\frac{\hbar c}{KE}\right)^2 \frac{1}{(1-\cos\theta)^2}$$

As the energy of the electrons is raised enough to make them an effective nuclear probe, a number of other effects become significant, and the scattering behaviour diverges from the Rutherford formula.

The probing electrons are relativistic, they produce significant nuclear recoil, and they interact via their magnetic moment as well as by their charge. When the magnetic moment and recoil are taken into account, the expression is called the Mott cross section:

Electron Rutherford Formula

Electron magnetic moment effect

$$\frac{d\sigma}{d\cos\theta} = \left(\frac{d\sigma}{d\cos\theta}\right)_R \frac{(1+\cos\theta)/2}{\left[1+\frac{(1-\cos\theta)KE}{Mc^2}\right]}$$

Nuclear recoil effect

A major period of investigation of nuclear size and structure occurred in the 1950's with the work of Robert Hofstadter and others who compared their high energy electron scattering results with the Mott cross section. The illustration below from Hofstadter's work shows the divergence from the Mott cross section which indicates that the electrons are penetrating the nucleus - departure from point-particle scattering is evidence of the structure of the nucleus.

This productive period of research with electron scattering built the picture of the nucleus as a spherical distribution of positive charge of essentially constant density, so that the nuclear radius could be modeled by the relationship:

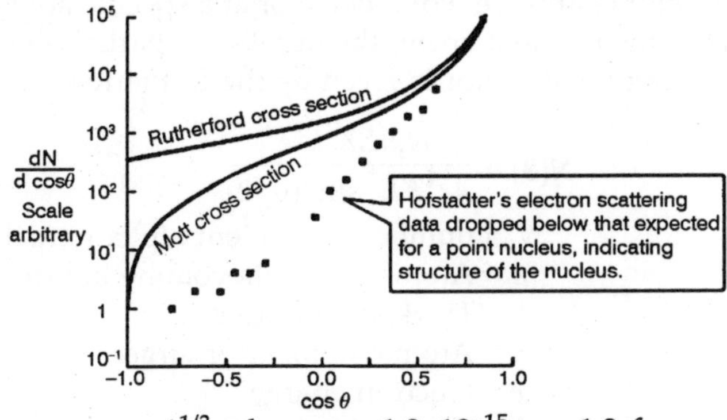

$$r = r_0 A^{1/3} \text{ where } r_0 = 1.2x10^{-15}\ m = 1.2\ fm$$

confirming the Fermi model for the nuclear radius.

Rutherford Scattering

Alpha particles from a radioactive source were allowed to strike a thin gold foil. Alpha particles produce a tiny, but visible flash of light when they strike a fluorescent screen. Surprisingly, alpha particles were found at large deflection angles and some were even found to be back-scattered.

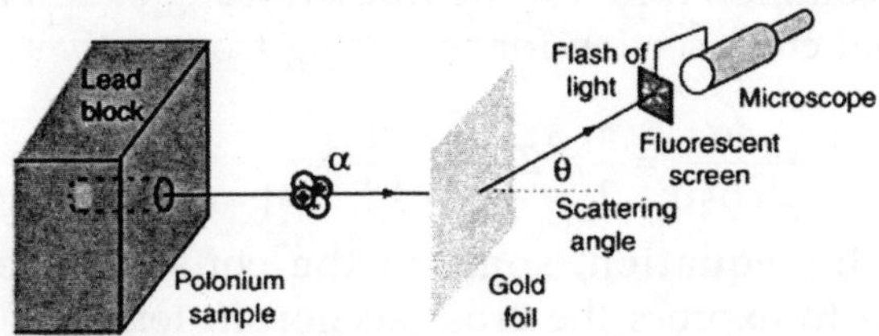

This experiment showed that the positive matter in atoms was concentrated in an incredibly small volume and gave birth to the idea of the nuclear atom. In so doing, it represented one of the great turning points in our understanding of nature.

If the gold foil were 1 micrometer thick, then using the diameter of the gold atom from the periodic table suggests that the foil is about 2800 atoms thick.

Rutherford Scattering Formula

The scattering of alpha particles from nuclei can be modeled from the Coulomb force and treated as an orbit. The scattering process can be treated statistically in terms of the cross-section for interaction with a nucleus which is considered to be a point charge Ze. For a detector at a specific angle with respect to the incident beam, the number of particles per unit area striking the detector is given by the Rutherford formula:

$$N(\theta) = \frac{N_i n L Z^2 k^2 e^4}{4r^2 KE^2 \sin^4(\theta/2)}$$

N_i = number of incident alpha particles
n = Atoms per units volume in target
L = Thickness of target
Z = Atomic number of target
e = Electron charge
k = Coulomb's constant
r = Target-to-detector distance
KE = Kinetic energy of alpha
θ = Scattering angle

The predicted variation of detected alphas with angle is followed closely by the Geiger-Marsden data. The above form includes the cross-section for scattering for a given nucleus and the nature of the scattering film to get the scattered fraction. Another common form for the Rutherford equation is just the differential cross section for scattering from a given nucleus.

$$\frac{d\sigma}{d\cos\theta} = \frac{\pi}{2} z^2 Z^2 \alpha^2 \left(\frac{\hbar c}{KE}\right)^2 \frac{1}{(1-\cos\theta)^2}$$

For this equation, some of the constants have been combined to express the cross section in terms of the fine-structure constant, α.

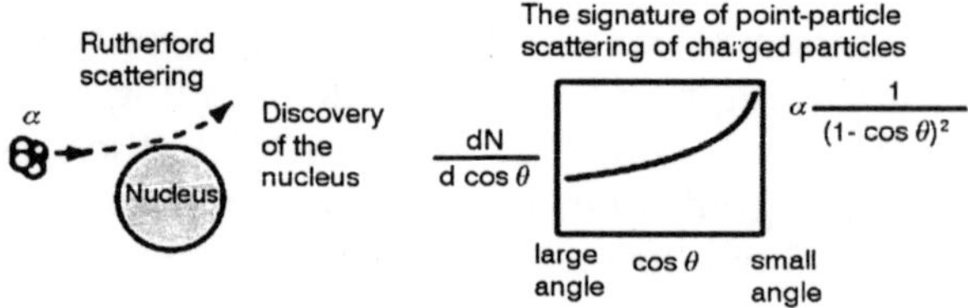

This form of the scattering formula serves as a signature for scattering off a point target in which no structure is evident.

The point of departure from Rutherford scattering in the case of the nucleus was the basis for the earliest evaluations of the nuclear radius. The departure from the point-particle form of scattering has been an indicator of nuclear structure and then at higher energies, the structure of the proton.

Water as Moderator

Neutrons from fission have very high speeds and must be slowed greatly by water "moderation" to maintain the chain reaction. The uranium-235 is enriched to 2.5 - 3.5% to allow ordinary water to be the moderator.

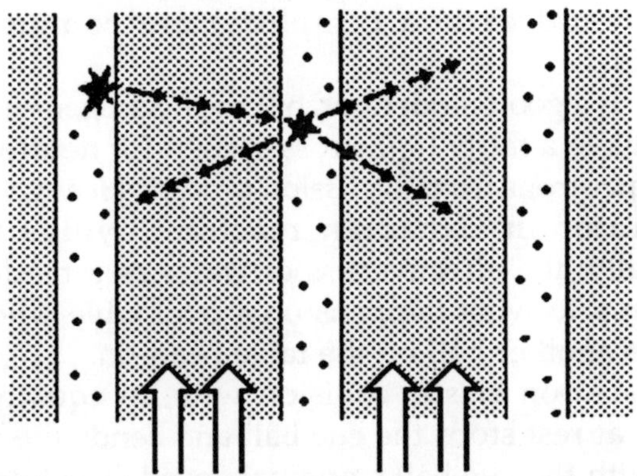

Loss or the water coolant kills the chain reactionsince the fuel configuration is not "critical" without water moderation.

Even with the moderator, the fuel is not "critical" without the inclusion of the "delayed" neutrons which may be emitted several minutes after the fissi

Moderation of Fast Neutrons

The neutrons emitted in nuclear fission reactions have high energies, typically in the range of 1 MeV. But the cross section for neutron capture leading to fission is greatest for neutrons of energy around 1 eV, a million times less.

Neutrons with energies less than one electron volt are commonly referred to as "thermal neutrons" since they have energies similar to what particles have as a result of ordinary room-temperature thermal energy. It is necessary to slow down the neutrons for efficient operation of a nuclear reactor, a process called moderation.

While neutrons are efficiently slowed by inelastic scattering from U-238, the non-fissionable isotope of uranium, when their energies are higher than 1 MeV, the remainder of the process of slowing them down must be done by elastic scattering from other nuclei.

When a neutron collides elastically with another nucleus at rest in the medium, it transfers some of its energy to it. The maximum transfer of energy occurs when the target nucleus is comparable in mass to the projectile.

Water and carbon (graphite) are commonly used moderators.

Water is a good moderator, but the hydrogens in the water molecule have a fairly high cross section for neutron capture, removing neutrons from the fission process. Heavy water, used as moderator in Canadian reactors, avoids this loss. Conceptually, the effectiveness of water as a moderator can be compared to what happens on a pool table when the cue ball strikes another ball on the table head-on.

The head-on elastic collision with an equally massive target ball at rest stops the cue ball and sends the target ball forward with the cue ball's original speed. The hydrogens in the water play the role of the target ball and are effective in dramatically slowing the fast neutrons, even when the collision is not head-on.

Another conceptual image which may help with understanding the need for moderation is the nature of a short putt on the green of a golf course. The original experiments in the laboratory of Otto Hahn in Germany tried unsuccessfully to get uranium to absorb neutrons by bombarding them with fast neutrons - ^{235}U just has a very small probability of absorbing fast neutrons.

But it has a high probability of absorbing slow ones. If your golf ball is a few centimeters from the hole, you don't get out your driver and hit it as hard as possible - it just will not go into the cup that way. But a gentle tap with your putter has a high probability of success. Moderation to slow the neutrons by collisions with nuclei of similar mass dramatically increases the probability of neutron capture leading to fission.

Critical Mass

For a chain reaction of nuclear fission, such as that of uranium-235, is to sustain itself, then at least one neutron from each fission must strike another U-235 nucleus and cause a fission. If this condition is just met, then the reaction is said to be "critical" and will continue. The mass of fissile material required to achieve this critical condition is said to be a critical mass. The critical mass depends upon the concentration of U-235 nuclei in the fuel material as well as its geometry.

As applied for the generation of electric energy in nuclear reactors, it also depends upon the moderation used to slow down the neutrons. In those reactors, the critical condition also depends upon neutrons from the fission fragments, called delayed neutrons. For weapons applications, the concentration U-235 must be much higher to create a condition called "prompt criticality". This means that it is critical with only the neutrons directly produced in the fission process. For U-235 enriched to "bomb-grade" uranium, the critical mass may be as small as about 15 kg in a bomb configuration.

Delayed Neutrons From Fission

One of the safety factors built into the nuclear reactors which are used for electricity generation is that they are only critical with the inclusion of the delayed neutrons which are emitted by some of the fission fragments. Some of these fragments emit neutrons as a part of their radioactive decay, and these neutrons can contribute to fission of any U-235 nucleus they strike.

A nuclear power reactor controls the fission chain reaction by moderating the neutrons and with the use of control rods which may be inserted in the reactor core to absorb neutrons and slow down the reaction. These control rods may be adjusted so that the reaction remains critical only with the inclusion of the delayed neutrons. About 0.65% of the neutrons are delayed by an average of 14 seconds, giving significant increase in the generation time and the time for reaction to an emergency in such a power reactor. Of course in a weapons application, these delayed neutrons are not significant, so weapons-grade uranium is enriched to over 90% U-235.

Fission Fragments

When uranium-235 undergoes fission, the average of the fragment mass is about 118, but very few fragments near that average are found. It is much more probable to break up into unequal fragments, and the most probable fragment masses are around mass 95 and 137. Most of these fission fragments are highly unstable (radioactive), and some of them such as cesium-137 and strontium-90 are extremely dangerous when released to the environment.

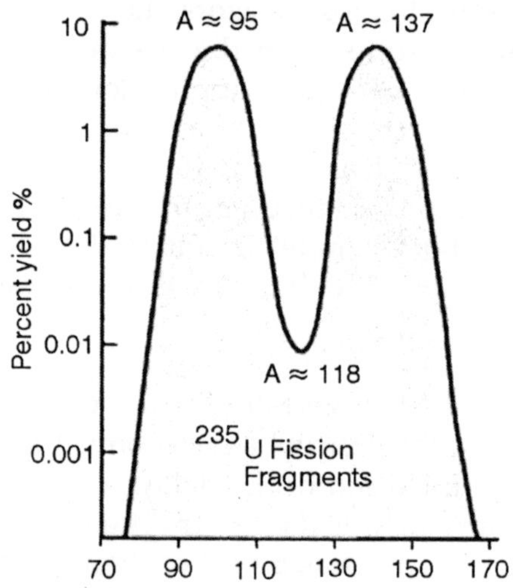

Fig. Mass number A of fission fragment

Fission Fragment Example

A common pair of fragments from uranium-235 fission is xenon and strontium:

$$^{235}U + n \rightarrow {}^{236}U^* \rightarrow {}^{140}Xe + {}^{94}Sr + 2n$$

Highly radioactive, the xenon decays with a half-life of 14 seconds and finally produces the stable isotope cerium-140.

Strontium-94 decays with a half-life of 75 seconds, finally producing the stable isotope zirconium-94. These fragments are not so dangerous as intermediate half-life fragments such as cesium-137.

Fission Fragment Decay

This particular set of fragments from uranium-235 fission undergoes a series of beta decays to form stable end products.

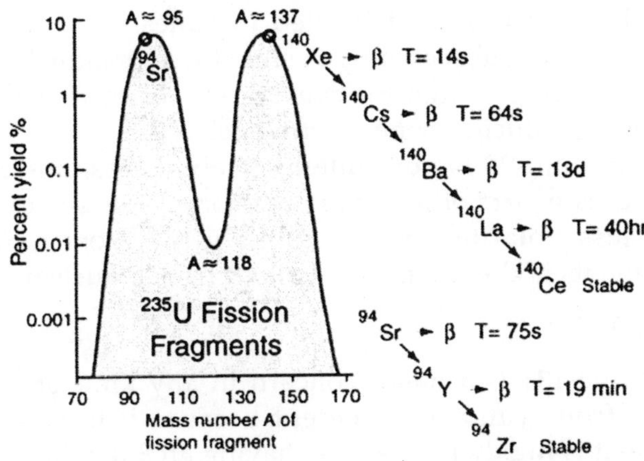

Cesium-137

Cesium-137 and strontium-90 are the most dangerous radioisotopes to the environment in terms of their long-term effects. Their intermediate half-lives of about 30 years suggests that they are not only highly radioactive but that they have a long enough halflife to be around for hundreds of years. Iodine-131 may give a higher initial dose, but its short halflife of 8 days ensures that it will soon be gone. Besides its persistence and high activity, cesium-137 has the further insidious property of being mistaken for potassium by living organisms and taken up as part of the fluid electrolytes. This means that it is passed on up the food chain and reconcentrated from the environment by that process.

Strontium-90

Strontium-90 and cesium-137 are the radioisotopes which

should be most closely gaurded against release into the environment. They both have intermediate halflives of around 30 years, which is the worst range for half-lives of radioactive contaminants. It ensures that they are not only highly radioactive but also have a long enough halflife to be around for hundreds of years.

Strontium-90 mimics the properties of calcium and is taken up by living organisms and made a part of their electrolytes as well as deposited in bones. As a part of the bones, it is not subsequently excreted like cesium-137 would be. It has the potential for causing cancer or damaging the rapidly reproducing bone marrow cells.

Strontium-90 is not quite as likely as cesium-137 to be released as a part of a nuclear reactor accident because it is much less volatile, but is probably the most dangerous components of the radioactive fallout from a nuclear weapon.

Iodine-131

Iodine-131 is a major concern in any kind of radiation release from a nuclear accident because it is volatile and because it is highly radioactive, having an 8 day half-life. It is of further concern in the human body because iodine is quickly swept up by the thyroid, so that the total intake of iodine becomes concentrated there.

The thyroid has a maximum uptake of iodine, however, so some protection against iodine releases can be afforded by taking potassium iodide tablets to load up the thyroid to capacity so that radioactive iodine would be more likely to be excreted.

Chapter 10

Nuclear Fusion

Nuclear energy can also be released by fusion of two light elements (elements with low atomic numbers). The power that fuels the sun and the stars is nuclear fusion. In a hydrogen bomb, two isotopes of hydrogen, deuterium and tritium are fused to form a nucleus of helium and a neutron. This fusion releases 17.6 MeV of energy. Unlike nuclear fission, there is no limit on the amount of the fusion that can occur.

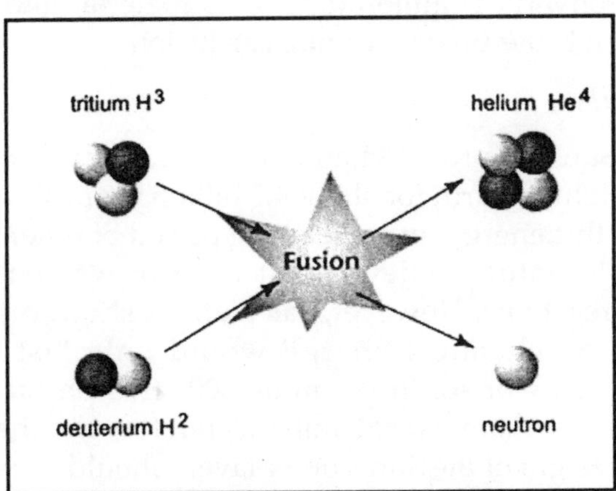

Fig. Nuclear Fusion

If light nuclei are forced together, they will fuse with a yield of energy because the mass of the combination will be less than the sum of the masses of the individual nuclei. If the combined nuclear mass is less than that of iron at the peak of the binding energy curve, then the nuclear particles will be more tightly bound than they were in the lighter nuclei, and that decrease in mass comes off in the form of energy according to the Einstein relationship. For elements heavier than iron, fission will yield energy. For potential nuclear energy sources for the Earth, the deuterium-tritium fusion reaction contained by some kind of magnetic confinement seems the most likely path. However, for the fueling of the stars, other fusion reactions will dominate.

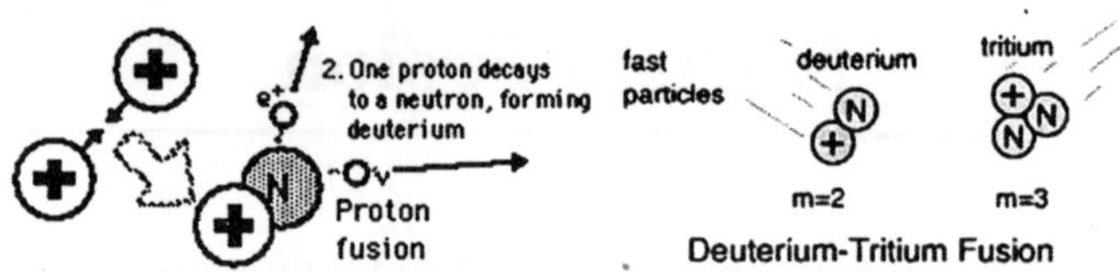

Quest. What makes the Sun shine?

Ans. The simple answer is that deep inside the core of the Sun, enough protons can collide into each other with enough speed that they stick together to form a helium nucleus and generate a tremendous amount of energy at the same time. This process is called nuclear fusion. Every second, a star like our Sun converts 4 million tons of its material into heat and light through the process of nuclear fusion.

The Details

Our Sun has provided an essentially constant amount of heat and light to Earth for about 4.5 billion years. But just what generates this energy for such a long period of time? Scientists of the 19th century believed that the Sun was powered by chemical reactions. However, calculations showed that a star powered on chemical energy would only last maybe a thousand years or so. In the mid-1800s two physicists, Lord Kelvin and Hermann von Helmholtz, put forward the idea that the huge weight of the Sun's outer layers should cause the Sun to gradually contract. As it contracts, the gases in its interior become compressed and when a gas is compressed its temperature increases.

Kelvin and Helmholtz argued that gravitational contraction would cause the Sun's gases to become hot enough to radiate heat energy into space. This process does in fact happen in the protostar phase of stellar formation. However, this kind of contraction cannot be the main source of stellar energy for billions of years...a hundred million, maybe, but not a billion. A clue to the source of stellar energy was provided by Albert Einstein. In 1905, while developing his special theory of relativity, Einstein showed that mass can be converted into energy and vice-versa. These quantities are related by the mass-energy relation

$$E = mc^2$$

where E is the energy released (in units called Joules) from the conversion of a mass m (in units of kg), and c is the speed of light (in meters per second). In 1920, British astronomer Arthur Eddington proposed that the Sun and other stars are powered by nuclear reactions.

Deuterium-Tritium Fusion

The most promising of the hydrogen fusion reactions which make up the deuterium cycle is the fusion of deuterium and tritium. The reaction yields 17.6 MeV of energy but requires a temperature of approximately 40 million Kelvins to overcome the coulomb barrier and ignite it. The deuterium fuel is abundant, but tritium must be either bred from lithium or gotten in the operation of the deuterium.

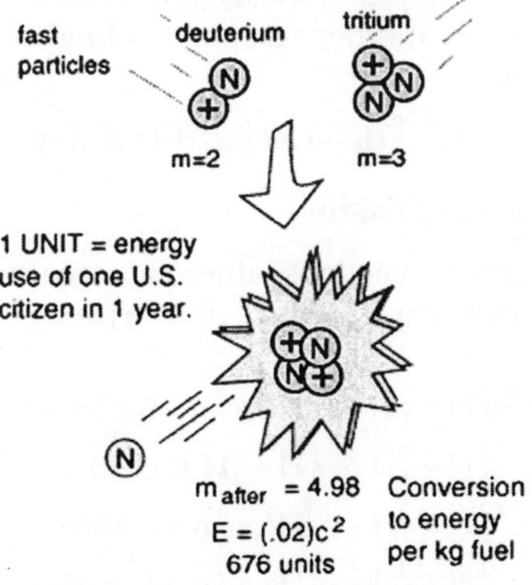

Hans Bethe realized that a proton smashing into another

proton with enough force could be the reaction to power the Sun. In 1938, Bethe and his colleagues presented a fully developed proton-proton chain of reactions that converts hydrogen into helium, which would allow the Sun to shine for about 10 billion years.

The proton-proton cycle, as it is now called, is known to be responsible for about 98% of the Sun's energy production in its inner core. Bethe won the 1967 Nobel Prize in Physics for his work concerning energy production in stars.

Hydrogen Fusion Reactions

Even though a lot of energy is required to overcome the Coulomb barrier and initiate hydrogen fusion, the energy yields are enough to encourage continued research. Hydrogen fusion on the earth could make use of the reactions:

$$\left.\begin{aligned} {}^2_1\text{H} + {}^2_1\text{H} &\rightarrow {}^3_2\text{He} + {}^1_0\text{n} + 3.27\ \text{MeV} \\ {}^2_1\text{H} + {}^2_1\text{H} &\rightarrow {}^3_1\text{H} + {}^1_1\text{H} + 4.03\ \text{MeV} \end{aligned}\right\}\text{deuterium-deuterium fusion}$$

$${}^2_1\text{H} + {}^3_1\text{H} \rightarrow {}^4_2\text{He} + {}^1_0\text{n} + 17.59\ \text{MeV}\ \text{deuterium-tritium fusion}$$

These reactions are more promising than the proton-proton fusion of the stars for potential energy sources. Of these the deuterium-tritium fusion appears to be the most promising and has been the subject of most experiments. In a deuterium-deuterium reactor, another reaction could also occur, creating a deuterium cycle:

$${}^2_1\text{H} + {}^3_2\text{He} \rightarrow \text{He}^4_2\text{H}^1_1 + 18.3\ \text{MeV}$$

Deuterium Cycle of Fusion

The four fusion reactions which can occur with deuterium can be considered to form a deuterium cycle. The four reactions:

$${}^2_1\text{H} + {}^2_1\text{H} \rightarrow {}^3_2\text{He} + {}^1_0\text{n} + 3.3\ \text{MeV}$$

$${}^2_1\text{H} + {}^2_1\text{H} \rightarrow {}^3_1\text{H} + {}^1_1\text{H} + 4.0\ \text{MeV}$$

$${}^2_1\text{H} + {}^2_1\text{H} \rightarrow {}^3_2\text{He} + {}^1_0\text{n} + 3.3\ \text{MeV}$$

$${}^2_1\text{H} + {}^3_1\text{H} \rightarrow {}^4_2\text{He} + {}^1_0\text{n} + 17.6\ \text{MeV}$$

can be combined as

$$6{}_1^2\text{H} + {}_1^3\text{H} + {}_2^3\text{He} \rightarrow 2{}_2^4\text{He} + {}_1^3\text{H} + 2{}_1^1\text{H} + 2{}_0^1\text{n} + 43.2\ \text{MeV}$$

or

$$6{}_1^2\text{H} \rightarrow 2{}_4^2\text{He} + 2{}_1^1\text{H} + 2{}_0^1\text{n} + 43.2\ \text{MeV}$$

Tritium Breeding

Deuterium-Tritium fusion is the most promising of the hydrogen fusion reactions, but no tritium occurs in nature since it has a 10 year half-life. The most promising source of tritium seems to be the breeding of tritium from lithium-6 by neutron bombardment with the reaction

$${}_3^6\text{Li} + {}_0^1\text{n} \rightarrow {}_2^4\text{He} + {}_1^3\text{H} + 4.8\ \text{MeV}$$

which can be achieved by slow neutrons. This would occur if lithium were used as the coolant and heat transfer medium around the reaction chamber of a fusion reactor.

Lithium-6 makes up 7.4% of natural lithium. While this constitutes a sizable supply, it is the limiting resource for the D-T process since the supply of deuterium fuel is virtually unlimited. With fast neutrons, tritium can be bred from the more abundant Li-7:

$${}_0^1\text{n(fast)} + {}_3^7\text{Li} \rightarrow {}_1^3\text{H} + {}_2^4\text{He} + {}_0^1\text{n(slow)}$$

Deuterium Source

Since the most practical nuclear fusion reaction for power generation seems to be the deuterium-tritium reaction, the sources of these fuels are important. The deuterium part of the fuel does not pose a great problem because about 1 part in 5000 of the hydrogen in seawater is deuterium. This amounts to over 10 × 15 tons of deuterium.

Viewed as a potential fuel for a fusion reactor, a gallon of seawater could produce as much energy as 300 gallons of gasoline. The tritium part of the fuel is more problematic - there is no sizable natural source since tritium is radioactive with a halflife of about 10 years. It would have to be obtained by breeding the tritium from lithium.

Nuclear Binding Energy

Nuclei are made up of protons and neutron, but the mass

of a nucleus is always less than the sum of the individual masses of the protons and neutrons which constitute it. The difference is a measure of the nuclear binding energy which holds the nucleus together. This binding energy can be calculated from the Einstein relationship:

$$\text{Nuclear binding energy} = \Delta mc^2$$

For the alpha particle Δm= 0.0304 u which gives a binding energy of 28.3 MeV.

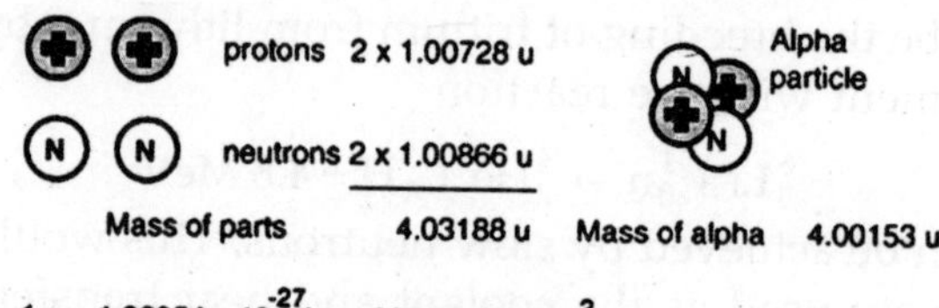

The enormity of the nuclear binding energy can perhaps be better appreciated by comparing it to the binding energy of an electron in an atom. The comparison of the alpha particle binding energy with the binding energy of the electron in a hydrogen atom is shown below. The nuclear binding energies are on the order of a million times greater than the electron binding energies of atoms.

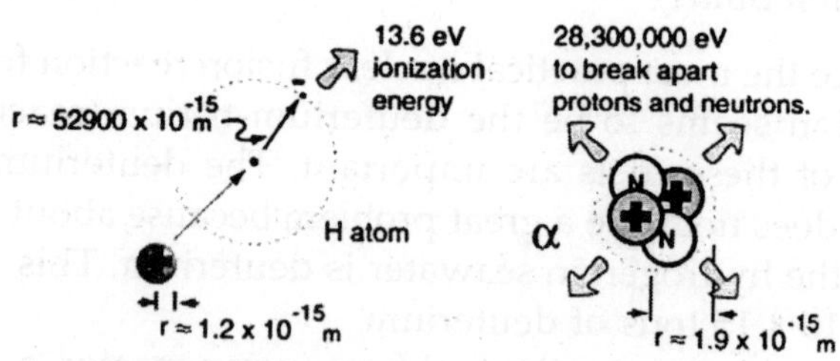

Comparison of atomic and nuclear scales and binding energy

Proton

Along with neutrons, protons make up the nucleus, held together by the strong force. The proton is a baryon and is considered to be composed of two up quarks and one down quark. It has long been considered to be a stable particle, but recent developments of grand unification models have suggested that it might decay with a half-life of about 10^{32} years. Experiments are underway to see if such decays can be detected. Decay of the proton would violate the conservation of baryon number, and in doing so would be the only known process in nature which does so.

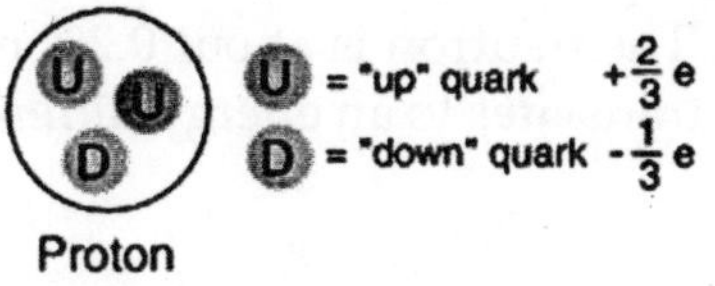

U = "up" quark $+\frac{2}{3}e$

D = "down" quark $-\frac{1}{3}e$

Proton

$m_p = 1836.15\ m_e$

Mass = 1.6726×10^{-27} kg

= 938.27231 MeV/c^2

= 1.00727647 u

When we say that a proton is made up of two up quarks and a down, we mean that its net appearance or net set of quantum numbers match that picture. The nature of quark confinement suggests that the quarks are surrounded by a cloud of gluons, and within the tiny volume of the proton other quark-antiquark pairs can be produced and then annihilated without changing the net external appearance of the proton.

Proton Decay

One of the implications of the grand unification theories is that the proton should decay with a half-life on the order of 10^{32} years. Such a long half-life is exceedingly difficult to measure, but the hope of doing led to a deep mine experiment in the Soudan iron mines of Minnesota. The Soudan 2 Proton Decay experiment ran from 1989-2001 without observing any convincing proton decays. Such experiments serve to push back the lower bound on the proton decay halflife. Another set of experiments designed to detect proton decay was carried out in the water Cherenkov detector at Super Kamiokande in Japan. Ed Kearns of Boston University suggested the following tentative bounds for proton decay the proton lifetime has been pushed out to 10^{33} years. The list of decay modes at left is not exhaustive. There are other modes under investigation that would suggest supersymmetry if found.

Neutron

Along with protons, neutrons make up the nucleus, held

together by the strong force. The neutron is a baryon and is considered to be composed of two down quarks and one up quark. A free neutron will decay with a half-life of about 10.3 minutes but it is stable if combined into a nucleus. The decay of the neutron involves the weak interaction as indicated in the Feynman diagram to the right. This fact is important in models of the early universe. The neutron is about 0.2% more massive than a proton, which translates to an energy difference of 1.29 MeV.

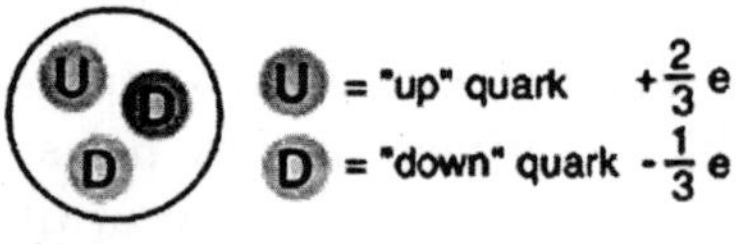

U D D
Neutron

U = "up" quark $+\frac{2}{3}e$
D = "down" quark $-\frac{1}{3}e$

$m_p = 1838.68\ m_e$

Mass = 1.6749×10^{-27} kg
= 939.5656 MeV/c^2
= 1.0086647 u

The decay of the neutron is associated with a quark transformation in which a down quark is converted to an up by the weak interaction. The average lifetime of 10.3 min/0.693 = 14.9 minutes is surprisingly long for a particle decay that yields 1.29 MeV of energy.

You could say that this decay is steeply "downhill" in energy and would be expected to proceed rapidly. It is possible for a proton to be transformed into a neutron, but you have to supply 1.29 MeV of energy to reach the threshold for that transformation.

In the very early stages of the big bang when the thermal energy was much greater than 1.29 MeV, we surmise that the transformation between protons and neutrons was proceeding freely in both directions so that there was an essentially equal population of protons and neutrons.

Decay of the Neutron

A free neutron will decay with a half-life of about 10.3 minutes but it is stable if combined into a nucleus. This decay is an example of beta decay with the emission of an electron and an electron antineutrino. The decay of the neutron involves the weak interaction as indicated in the Feynman diagram to the right.

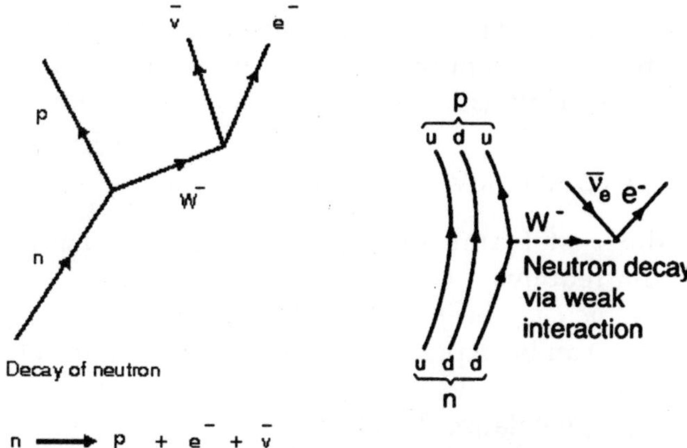

$n \longrightarrow p + e^- + \bar{v}$

The neutron's decay identifies it as the transformation of one of the neutron's down quarks into an up quark. It is an example of the kind of quark transformations that are involved in many nuclear processes, including beta decay.

The decay of the neutron is a good example of the observations which led to the discovery of the neutrino. An analysis of the energetics of the decay can be used to illustrate the dilemmas which faced early investigators of this process.

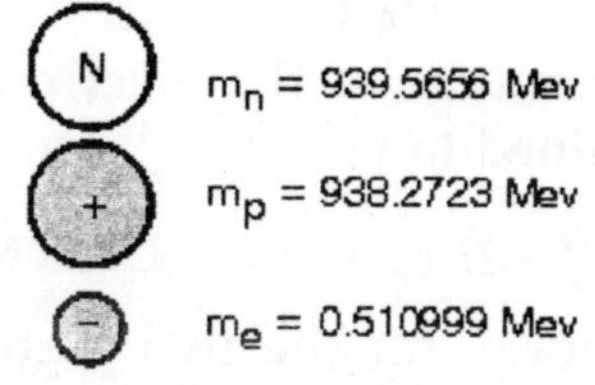

0.7823 Mev = Q for $n \rightarrow p + e^-$

Using the concept of binding energy, and representing the masses of the particles by their rest mass energies, the energy yield from neutron decay can be calculated from the particle masses. The energy yield is traditionally represented by the symbol Q. Because energy and momentum must be conserved in the decay, it will be shown that the lighter electron will carry away most of the kinetic energy.

With a kinetic energy of this magnitude, the relativistic kinetic energy expression must be used. For the moment we presume (incorrectly) that the decay involves just the proton and electron as products. The energy yield Q would then be divided between the proton and electron. The electron will get most of the kinetic energy and will be relativistic, but the proton is non-relativistic. The energy balance is then

$$Q = 0.7823 \text{ MeV} = KE_e + \frac{1}{2}m_p v_p^2 = KE_e + \frac{p_p^2 c^2}{2m_p c^2}$$

In the rest frame of the neutron, conservation of momentum requires

$$pc_{\text{electron}} = -pc_{\text{proton}}$$

and pc_{electron} can be expressed in terms of the electron kinetic energy

The energy balance then becomes

$$0.7823 \text{ MeV} = KE_e = \frac{KE_e^2 + 2KE_e \cdot m_e c^2}{2m_p c^2}$$

When you substitute the numbers for this value of Q, you see that the $KE_e{}^2$ term is negligible, so the required kinetic energy of the electron can be calculated.The required electron kinetic energy for this two-particle decay scheme is

$$0.7823 \text{ MeV} = KE_e \left[1 + \frac{m_e}{m_p}\right]; \; KE_e = 0.7819 \text{ MeV}$$

Likewise, the momentum of the electron for this two particle decay is constrained to be

$$pc_{\text{electron}} = \sqrt{KE_e^2 + 2KE_e \cdot m_e c^2} = 1.188 \text{ MeV}$$

Momentum and energy for the two-particle decay are constrained to these values, but this is not the way nature behaves. The observed momentum and energy distributions for the electron are as shown below.

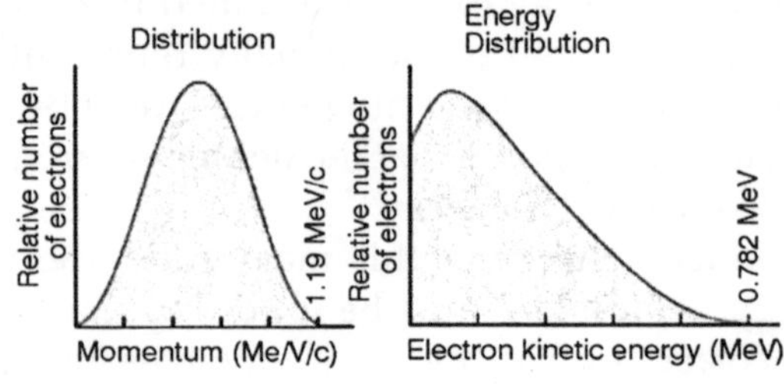

The fact that the electrons produced from the neutron decay had continuous distributions of energy and momentum was a clear indication that there was another particle emitted along with the electron and proton. It had to be a neutral particle and in certain decays carried almost all the energy and momentum of the decay.

This would not have been so extraordinary except for the fact that when the electron had its maximum kinetic energy, it accounted for all the energy Q available for the decay. So there was no energy left over to account for the mass energy of the other emitted particle. The early experimenters were faced with the dilemma of a particle which could carry nearly all the energy and momentum of the decay but which had no charge and apparently no mass!

The mysterious particle was called a neutrino, but it was twenty five years before unambiguous experimental observation of the neutrino was made by Cowan and Reines. The present understanding of the decay of the neutron is

$$n \rightarrow p + e^- + \underset{\text{Electron Antineutrion}}{\overline{\nu}_e}$$

This decay illustrates some of the conservation laws which govern particle decays. The proton in the product satisfies the conservation of baryon number, but the emergence of the electron unaccompanied would violate conservation of lepton number.

The third particle must be an electron antineutrino to allow the decay to satisfy lepton number conservation. The electron has lepton number 1, and the antineutrino has lepton number –1.

The Strong Force

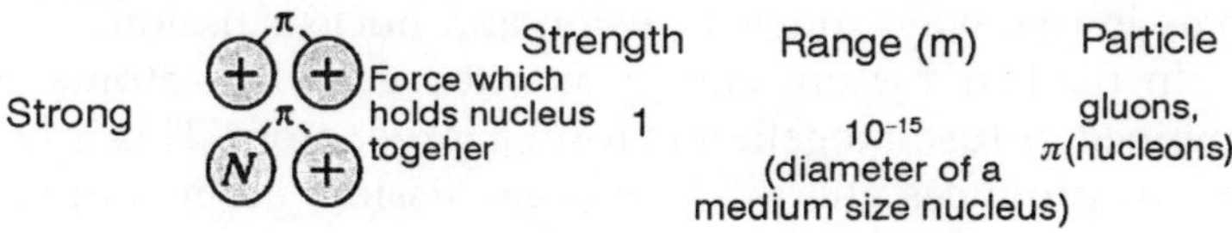

A force which can hold a nucleus together against the enormous forces of repulsion of the protons is strong indeed. However, it is not an inverse square force like the electromagnetic force and it has a very short range. Yukawa modeled the strong force as an exchange force in which the exchange particles are pions and other heavier particles. The range of a particle exchange force is limited by the uncertainty principle. It is the strongest of the four fundamental forces.

Since the protons and neutrons which make up the nucleus are themselves considered to be made up of quarks, and the quarks are considered to be held together by the colour force, the strong force between nucleons may be considered to be a residual colour force.

In the standard model, therefore, the basic exchange particle is the gluon which mediates the forces between quarks. Since the individual gluons and quarks are contained within the proton or neutron, the masses attributed to them cannot be used in the range relationship to predict the range of the force. When something is viewed as emerging from a proton or neutron, then it must be at least a quark-antiquark pair, so it is then plausible that the pion as the lightest meson should serve as a predictor of the maximum range of the strong force between nucleons.

The sketch is an attempt to show one of many forms the gluon interaction between nucleons could take, this one involving up-antiup pair production and annililation and producing a ð⁻ bridging the nucleons.

NUCLEAR ENERGY IS ENERGY FROM ATOMS

Nuclear energy is energy in the nucleus (core) of an atom. Atoms are tiny particles that make up every object in the universe. There is enormous energy in the bonds that hold atoms together. Nuclear energy can be used to make electricity. But first the energy must be released. It can be released from atoms in two ways: nuclear fusion and nuclear fission.

In nuclear fusion, energy is released when atoms are combined or fused together to form a larger atom. This is how the sun produces energy. In nuclear fission, atoms are split apart to form smaller atoms, releasing energy. Nuclear power plants use nuclear fission to produce electricity.

NUCLEAR FUEL - URANIUM

The fuel most widely used by nuclear plants for nuclear fission is uranium. Uranium is nonrenewable, though it is a common metal found in rocks all over the world. Nuclear plants use a certain kind of uranium, U-235, as fuel because its atoms are easily split apart. Though uranium is quite common, about 100 times more common than silver, U-235 is relatively rare. Most U.S. uranium is mined, in the Western United States. Once uranium is mined the U-235 must be extracted and processed before it can be used as a fuel.

During nuclear fission, a small particle called a neutron hits the uranium atom and splits it, releasing a great amount of energy as heat and radiation. More neutrons are also released. These neutrons go on to bombard other uranium atoms, and the process repeats itself over and over again. This is called a chain reaction.

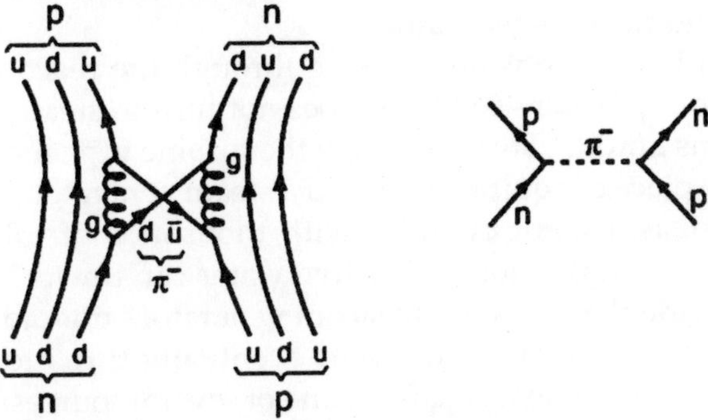

Nuclear Energy is Energy from Atoms

Most power plants burn fuel to produce electricity, but not nuclear power plants. Instead, nuclear plants use the heat given off during fission as fuel. Fission takes place inside the reactor of a nuclear power plant. At the centre of the reactor is the core, which contains the uranium fuel.

The uranium fuel is formed into ceramic pellets. The pellets are about the size of your fingertip, but each one produces the same amount of energy as 150 gallons of oil. These energy-rich pellets are stacked end-to-end in 12-foot metal fuel rods. A bundle of fuel rods is called a fuel assembly.

Fission generates heat in a reactor just as coal generates heat in a boiler. The heat is used to boil water into steam. The steam turns huge turbine blades. As they turn, they drive generators that make electricity. Afterward, the steam is changed back into water and cooled in a separate structure at the power plant called a cooling tower. The water can be used again and again.

Types of Reactors

Just as there are different approaches to designing and building airplanes and automobiles, engineers have developed different types of nuclear power plants. Two types are used in the United States: boiling-water reactors (BWRs), and pressurized-water reactors (PWRs).

In the BWR, the water heated by the reactor core turns directly into steam in the reactor vessel and is then used to power the turbine-generator.

In a PWR, the water passing through the reactor core is kept under pressure so that it does not turn to steam at all – it remains liquid. Steam to drive the turbine is generated in a separate piece of equipment called a steam generator. A steam generator is a giant cylinder with thousands of tubes in it through which the hot radioactive water can flow.

Outside the tubes in the steam generator, nonradioactive water (or clean water) boils and eventually turns to steam. The clean water may come from one of several sources: oceans, lakes or rivers. The radioactive water flows back to the reactor core, where it is reheated, only to flow back to the steam generator. Nuclear reactors are basically machines that contain and control chain reactions, while releasing heat at a controlled rate. In electric power plants, the reactors supply the heat to turn water into steam, which drives the turbine-generators. The electricity travels through high voltage transmission lines and low voltage distribution lines to homes, schools, hospitals, factories, office buildings, rail systems and other users.

NUCLEAR POWER AND THE ENVIRONMENT

Compared to electricity generated by burning fossil fuels, nuclear energy is clean. Nuclear power plants produce no air pollution or carbon dioxide but a small amount of emissions result from processing the uranium that is used in nuclear reactors. Like all industrial processes, nuclear power generation has by-product wastes: spent (used) fuels, other radioactive waste, and heat. Spent fuels and other radioactive wastes are the principal environmental concern for nuclear power. Most nuclear waste is low-level radioactive waste.

It consists of ordinary tools, protective clothing, wiping cloths and disposable items that have been contaminated with small amounts of radioactive dust or particles. These materials are subject to special regulation that govern their disposal so they will not come in contact with the outside environment.

THE PROS AND CONS OF NUCLEAR ENERGY

Pros

Little Pollution: As demand for electricity soars, the pollution produced from fossil fuel-burning plants is heading towards dangerous levels. Coal, gas and oil burning power plants are already responsible for half of America's air pollution. Burning coal produces carbon dioxide, which depletes the protection of the ozone. The soft coal, which many power plants burn, contains sulfur When the gaseous byproducts are absorbed in clouds, precipitation becomes sulfuric acid.. Coal also contains radioactive material. A coal-fired power plant emits more radiation into the air than a nuclear power plant.

The world's reserves of fossil fuels are running out. The sulfurous coal which many plants use is more polluting than the coal that was previously used. Most of the anthracite, which plants also burn, has been used up. As more soft coal is used, the amount of pollution will increase. According to estimates, fossil fuels will be burned up within fifty years. There are large reserves of uranium, and new breeder reactors can produce more fuel than they use. Unfortunately this doesn't mean we can have an endless supply of fuel Breeder reactors need a feedstock of uranium and thorium, so when we run out of these two fuels (in about 1000 years), breeder reactors will cease to be useful. This is still a more lengthy solution to the current burning of coal, gas, and oil.

Reliability: Nuclear power plants need little fuel, so they are less vulnerable to shortages because of strikes or natural disasters. International relations will have little effect on the supply of fuel to the reactors because uranium is evenly deposited around the globe.

One disadvantage of uranium mining is that it leaves the residues from chemical processing of the ore, which leads to radon exposure to the public. These effects do not outweigh the benefits by the fact that mining uranium out of the ground reduces future radon exposures. Coal burning leaves ashes that will increase future radon exposures. The estimates of radon show that it is safer to use nuclear fuel than burn coal. Mining of the fuel required to operate a nuclear plant for one year will avert a few hundred deaths, while the ashes from a coal-burning plant will cause 30 deaths.

Safety: Safety is both a pro and con, depending on which way you see it. The results of a compromised reactor core can be disastrous, but the precautions that prevent this from happening prevent it well. Nuclear power is one the safest methods of producing energy. Each year, 10,000 to 50,000 Americans die from respiratory diseases due to the burning of coal, and 300 are killed in mining and transportation accidents. In contrast, no Americans have died or been seriously injured because of a reactor accident or radiation exposure from American nuclear power plants.

There are a number of safety mechanisms that make the chances of reactor accidents very low. A series of barriers separates the radiation and heat of the reactor core from the outside. The reactor core is contained within a 9-inch thick steel pressure vessel. The pressure vessel is surrounded by a thick concrete wall. This is inside a sealed steel containment structure, which itself is inside a steel-reinforced concrete dome four feet thick. The dome is designed to withstand extremes such as earthquakes or a direct hit by a crashing airliner. There is also a large number of sensors that pick up increases in radiation or humidity. An increase in radiation or humidity could mean there is a leak. There are systems that control and stop the chain reaction if necessary. An Emergency Core Cooling System ensures that in the event of an accident there is enough cooling water to cool the reactor.

Cons

Meltdowns: If there is a loss of coolant water in a fission reactor, the rods would overheat. The rods that contain the uranium fuel pellets would dissolve, leaving the fuel exposed. The temperature would increase with the lack of a cooling source. When the fuel rods heat to 2800°C, the fuel would melt, and a white-hot molten mass would melt its way through the containment vessels to the ground below it. This is a worst case scenario, as there are many precautions taken to avoid this. Emergency water reservoirs are designed to immediately flood the core in the case of sudden loss of coolant. There are normally multiple sources of water to draw from, as the low pressure injection pumps, containment spray system, and refueling pumps are all potentially available, and all draw water from different sources.

The disaster at Three Mile Island was classified as a partial meltdown, caused by the failure to supply coolant to the core. Although the core was completely destroyed, the radioactive mass never penetrated the steel outlining the containment structure. Several feet of special concrete, a standard precaution, was capable of preventing leakage for several hours, giving operators enough time to fix the flooding system of the reactor core. The worst case of a nuclear disaster was in 1986 at the Chernobyl facility in the Ukraine.

A fire ripped apart the casing of the core, releasing radioactive isotopes into the atmosphere. Thirty-one people died as an immediate result. And estimated 15,000 more died in the surrounding area after exposure to the radiation. Three Mile Island and Chernobyl are just examples of the serious problems that meltdowns can create.

Radiation: Radiation doses of about 200 rems cause radiation sickness, but only if this large amount of radiation is received all at once. The average person receives about 200 millirems a year from everyday objects and outer space. This is referred to as background radiation. If all our power came from nuclear plants we would receive an extra 2/10 of a millirem a year.

The three major effects of radiation are nearly untraceable at levels below about 50 rems. In a study of 100,000 survivors of the atomic bombs dropped on Hiroshima and Nagasaki, there have been 400 more cancer deaths than normal, and there is not an above average rate of genetic disease in their children. During the accident at Three Mile Island in America, people living within a 50 mile radius only received an extra 3/10 of one percent of their average annual radiation. This was because of the containment structures, the majority of which were not breached. The containment building and primary pressure vessel remained undamaged, fulfilling their function.

Waste Disposal

The byproducts of the fissioning of uranium-235 remains radioactive for thousands of years, requiring safe disposal away from society until they lose their significant radiation values. Many underground sites have been constructed, only to be filled within months.

Storage facilities are not sufficient to store the world's nuclear waste, which limits the amount of nuclear fuel that can be used per year. Transportation of the waste is risky, as many unknown variables may affect the containment vessels. If one of these vessels were compromised, the results may be deadly.

Index

N

O

P

Q

R

S

T

U

V

W